5 February 74

For Mary and Jim Semans,
who, I hope, will love this bit
of Americana, as does
Morris Fradin

HEY-EY-EY, LOCK!

by

Morris Fradin

"For all who love the C & O Canal —
the unique link with our American heritage."

Printed in United States of America.

First U.S. Edition 1974
Library of Congress Catalog Card Number: 73-91394

see·and·know press
P.O. BOX 226 · CABIN JOHN MARYLAND · 20731

Introduction

BY "MR. C & O CANAL."

I was a barefoot mule boy, or hoggee, on the towpath of the Chesapeake and Ohio Canal long before 1908, when Abby and Mike barged from Georgetown, D.C. to Cumberland, Maryland, on the *Daisy Mae*.

I was born in 1894. Up-canal from Lock 44 in Williamsport. Within sight of the Conococheague Creek Aqueduct that Mosby's Raiders blasted during the Civil War. I've walked and worked up and down the C & O Canal, or lived near it, most of my lifetime.

Reading *Hey-ey-ey, Lock!*, by Morris Fradin, brings back vivid memories of those bygone days. Days and nights of adventure along a waterway rich in history.

Although many places and things mentioned in this book no longer exist, the reader may still visit Washington and quaint Georgetown, the Clara Barton House, Glen Echo Park, the Cabin John Bridge, Great Falls Tavern and its fine museum, Seneca, Harpers Ferry, Antietam, Sharpsburg, the Paw Paw Tunnel, and interesting old towns, aqueducts and dams, locks and lockhouses, still visible near or on the canal.

Because of *Hey-ey-ey, Lock!*, youngsters and oldsters from 'round the world will be stimulated to greater interest in the entire 185-mile length of the canal. It is a living nature center filled with beauty and serenity. And a wide range of recreational fun. Now a United States National Historical Park, it is part of our American heritage. And it's waiting for *you* to experience it.

It was indeed a happy day when I met Morris Fradin, for he has certainly reinspired my lifelong love affair with the C & O Canal and the Potomac River. I know his book will weave the same magic spell over you, too — even if he never drove mules on the C & O Canal as I did.

Williamsport, Maryland
October 1973

George W. "Hooper" Wolfe

CONTENTS

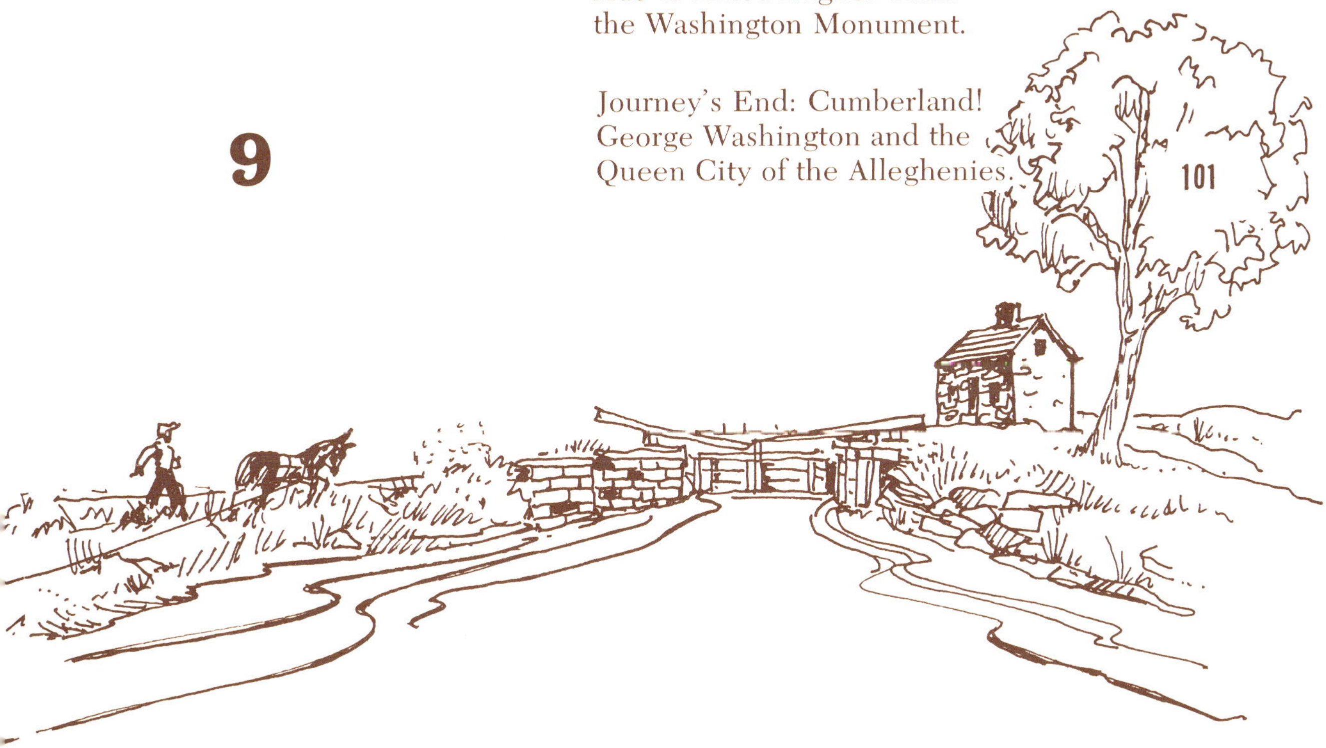

journey into history .....

DRY GOODS
GEORGETOWN
DRY GOODS
DAISY MAE

CHAPTER 1

"Westward, Ho!"
The Journey Begins.

"Hey-ey-ey, Lock!" shouted Captain Tom Clark, standing at the tiller of the *Daisy Mae*. "Lock ready! Lock ready!"

His daughter, Abigail, and her English cousin, Michael Walsh, sat on the bow of the canal barge, eating apples. Their bare legs dangled over the side; their heels drummed happily against the hull.

Mrs. Clark came up behind the children, tying the strings of a red-checked sunbonnet under her chin.

Just then the captain cried: "Quick, kids! Off the bow. Hit the deck."

"Ya-hoo!" yelled Abigail and gripped Michael's elbow. Mrs. Clark stepped back, but not fast enough. For the youngsters somersaulted backward and came to rest on her shoes and caught in the hem of her long skirt. She tried to back away, but tripped and tumbled down into a squirming heap. It finally untangled itself in a surge of giggles and laughter.

"My goodness!" gasped Mrs. Clark. She hitched her skirt higher and straightened it out. "How can anyone maintain dignity around here with a tomboy in the family? But you're certainly making Michael less shy. Now you come with me, Abigail. I have some chores for you."

"Yes, mother," said the girl, "Coming." She turned to nudge her cousin. She stood a head taller than Michael and spoke like an older sister.

"Watch old John on the towpath," she said, "so you'll be a good hoggee when you spell him in mule-driving."

"Righto!" said Michael. He watched black Mr. Manson on the tow-path, leading Bozo the mule up the ramp of the lock.

Clippedy-clop! clippedy-clop! went Bozo's hooves. Jingle! Jingle! went the bells on his harness. He strained into his collar and the long towrope, tied to the boat's bow, leaped, dripping and taut, into the air.

The great wooden gates of Lock 3 of the Chesapeake and Ohio Canal, in Georgetown, were spread open in welcome. Old John and the lock-keeper slipped the towrope around a snubbing post to help ease the *Daisy Mae* between the stone walls without scraping the lock or the hull. The boat glided to a gentle stop in the lock chamber.

Up front, another pair of massive, wooden gates was already shut tight. Forming a wall, they held back the water at a higher level.

Michael looked up, awed, at the black, pitch-covered, gates, and at the tops of the lock walls, where Old John and the lockkeeper stood peering down on him. Michael turned and looked toward the stern of the *Daisy Mae,* at the view framed by the lock and the open gates.

"It's beautiful . . . " he breathed. "Beautiful . . . like a picture in a storybook. Like...like a canal village back home in England."

And, indeed, the cobblestoned Thomas Jefferson Street, the red brick sidewalks, the black iron lampposts, topped by glass globes over gas mantles, the small, old rowhouses snuggling against each other, their doors gleaming with brass doorknobs and knockers, did resemble a part of the Old World.

Once a seaport, crowded with full-rigged merchantmen from 'round the world, Georgetown was already 25 years old when the United States was "born" in far-off Philadelphia. Georgetown was a bustling port, when Washington City was mostly swampland. And its Custom House still collected duties on foreign goods consigned to the District of Columbia, or exported from there to distant lands.

The downstream gates of the lock creaked as the men put their weight against the balance beams. The gates came together slowly, shutting off Michael's view . . .

Only the other day, Michael Walsh, escorted by a ship's steward, had stepped down the gangplank of the Cunard liner from Liverpool, onto the pier at Baltimore. "This New World is such a busy place," he thought, looking about him, curiously. The youth was identified by a large tag, secured to the left lapel of his jacket, and to each carpetbag carried by the steward.

And there, at the bottom of the gangplank, waited his tall uncle, wearing the peaked officer's hat of a mariner. His bushy eyebrows framed alert gray eyes. His broad, smiling face, flanked by long, wide sideburns, broke into a smile of greeting.

"Michael! Welcome to America," he boomed. He removed his hat and bent to embrace his nephew for the first time. "Recognized you from the photograph your mother sent me." He held the boy back, examining his dark hair, dark eyes, ruddy cheeks, and freckled nose that turned upward to end in a stub. He immediately liked the lad's cheery appearance.

"Welcome to America," he repeated, smiling broadly.

And Michael had responded with a firm handshake. "Happy to meet you, sir — I mean, Uncle Thomas. I also knew you from this photograph mother gave me."

Captain Clark examined the picture clutched in the boy's hand. "A reasonable likeness," he said, snapping a salute at his nephew. "At your service, sir!" He took the two bags from the waiting steward and pressed a coin into his hand. Then, with a smiling jerk of the head, the captain strode off and Michael scampered to keep up with his long legs.

They rode an electric trolley car to the railroad station. There, they boarded a train coach and settled themselves comfortably on the fuzzy red seats next to an open window.

"All aboard!" cried the conductor.

Whistles shrilled, the locomotive bell clanged, and the engine went chuff! chuff! . . . chuff! chuff! and drew slowly out of the station, huffing and puffing black clouds of smoke and white steam. The great drive wheels turned and screeched over the rails. Soot started to settle over the seats and over everyone's clothes, but the car was too stuffy to shut the windows just yet.

Captain Clark surveyed his nephew quietly for some moments.

"So my sister finally allowed you to slip her apron strings this summer," he said, and his bushy eyebrows bobbed up and down as he spoke. He pulled an envelope from his inside coat pocket and tapped it with his forefinger. "Off to the Continent with your father — she wrote me. Well, Michael," he patted the lad's arm, and a smile covered the captain's friendly face, "we'll take good care of you in America."

"Thank you, sir," said Michael, politely. "Thank you very much."

Opening the letter and glancing over its contents, the captain mused: "You're an unusual lad; your mother writes that you like history. She asks that I tell you and show you all I can about your American heritage. Especially about the people and places along the Chesapeake and Ohio Canal where we live." He paused and stuffed the letter back in his coat. "So . . . I've tried to jam my head with answers to your every 'Why?', 'What's that?', and 'What-happened-there?' "

Then he had grinned and added: "I hope!"

A boy came through the car just then, carrying a wicker basket filled with apples. Captain Clark chose and paid for two handsome red-skinned ones. He polished them on his coat sleeve 'til they glistened, then handed one to Michael. The youth bit into it happily and the juice shot out and drooled down his chin. His uncle drew a pocketknife and cut his fruit into small slices. He peeled and popped them into his mouth, chewing with smacks of pleasure.

Michael looked out the window, watching the swirls of smoke from the locomotive drift over the farms that blurred by. Farmers in the fields stopped their work to wave and smile as the train clattered past. The engineer blasted the air with a long warning whistle, when the train neared crossings and clanged on the bell, as though he wanted everyone to see how proud he was of his swift and fiery engine.

The cars bounced and jiggled and swayed from side to side; they creaked and screamed around the curves, so that the brass kerosene lamps hanging from the curved ceilings swung in the opposite direction of the car's movements.

Uncle Thomas had dozed off in his seat and his head, like a pendulum, rolled gently from one shoulder to the other.

As the car wheels clickety-clacked toward Washington City, Michael relaxed in his seat, recalling what his mother had told him about Uncle Thomas:

"He's lovable, with a good sense of humor, and just chockfull of interesting tidbits of history. Sprinkles them, like seasoning, through his conversations."

Long before, Mrs. Walsh had told Michael that Uncle Thomas had been a Navy officer aboard the *USS Maine* when it blew up in Havana Harbor in 1898, bringing on the Spanish-American War.

Michael's mother had related that Uncle Thomas helped entertain Miss Clara Barton, founder of the American Red Cross, aboard the warship, only the day before it blew up in Cuba. "She visited your uncle in the hospital where he convalesced," said Mrs. Walsh, "and although Miss Barton was much older, the two became good friends. They discussed the Civil War, Red Cross history, and Miss Barton's relief and rescue work.

"Uncle Thomas was lucky to escape with only smoke inhalation damage to his lungs, when so many of his shipmates were killed outright or maimed.

"He returned to his former duty as a history instructor at the U.S. Naval Academy at Annapolis. But he never forgot Miss Barton's descriptions of her headquarters and home in Maryland, above the C & O Canal: how quiet it was there, among the trees, the birds, the wildlife. And how healthy he would be, living on the canal with his wife and infant daughter — your cousin, Abigail.

"So — when his doctors ordered Uncle Thomas to live outdoors, away from stuffy classrooms, he bought an old canal boat and renamed it the *Daisy Mae* for his wife.

"Your uncle and aunt now live on that boat, while Abigail attends a boarding school in Georgetown. She spends her summer vacations on the canal with her parents. She's a tomboy, I'm told, but very friendly and full of fun."

Michael watched the scenery slow up. The train was coming to a stop.

"Washington!" called the conductor, striding down the aisle. "Washington!" He shook awake the regular passengers he knew.

The engine came to a hissing, jolting stop, jarring Captain Clark awake. His eyes snapped open in a startled expression. He grabbed Michael's arm and bags and led the boy out of the train shed to an electric streetcar. It threaded its way through the streets of the National Capital, past the dignified Capitol and the impressive Library of Congress and up Pennsylvania Avenue, past the United States Post Office Department. Uncle Thomas pointed out the Washington Monument in the distance. Then they passed the Treasury Department and the White House, where President Theodore Roosevelt had his office and home, and the State, War, and Navy Building, next to it, on 17th Street.

The car clattered on. The motorman clanged the bell under his foot at every intersection, to keep pedestrians away from the net of the "cowcatcher" on the front of his car.

Eventually, the little "Toonerville trolley" rattled its way into Georgetown. And it was almost twilight when the tired travelers finally stepped aboard the *Daisy Mae.*

"Welcome! Welcome!" Mrs. Clark beamed, drying her hands on her gingham apron, before grasping Michael's outstretched hand. Then she hugged and kissed him, crushing the red-checked sunbonnet she wore over her small, pleasant face against his cheek.

"I say, madam," the captain tapped his wife gently on the shoulder, "What about me?" He pretended to pout.

"Oh, you!" his wife laughed, giving him a hearty kiss. She turned back to Michael, holding him off and admiring him. "My, you're a big lad. And a right handsome one, too," she added. "We'll have to get you something to eat; I know you must be starved. Boys always are. You can unpack your bags later." Turning, she called: "Abigail! Come meet your cousin from England."

A slim girl in a blue calico dress, with braided blonde pigtails flapping behind her merry, round face, came bounding, barefooted, from the cabin at the stern of the barge. Some strands of hair hung loose from her pigtails and she scratched at mosquito bites on her left arm.

"Hiya, cousin," she said, her blue eyes sparkling. She seized, gripped, and shook Michael's arm up and down, as though it were a pump handle. He winced, because of his pained fingers.

"Call me Abby, cousin, and I'll call you Mike. Okay?"

"Righto! I'd like that." he had managed to gasp. And behind his back he rubbed his bruised hand and thought: "Mother was right . . what a tomboy . . .!"

A blast signal from a tin horn, sounded by an impatient steersman on a boat behind the *Daisy Mae*, shattered Mike's thoughts of the past few days and made him aware of the present . . .

Old John was hurrying to help close the lock chamber on one downstream end. He grasped a long balance beam that extended from the wooden gate behind the barge. He pulled, digging his heels into the dirt of the towpath. The lockkeeper strained likewise on the opposite or berm wall. Slowly, the two heavy gate edges came together; they creaked shut behind the "locked-in" barge. It was afloat at the bottom of a great stone-and-wood box filled with water.

Plumes of water spurted through cracks in the closed front gate timbers; some water streamed from cracks in the stone walls, onto the top of the mule stable at the bow. Leftover water, from the canal above the lock, now frothed through a flume that half-circled the sides and rear of the lockhouse; it cascaded over the rocks at the lower end of the canal lock. The flumes were safety-valves: they prevented flooding of the land next to the lockhouses.

Another rude toot sounded from the horn of the barge behind the *Daisy Mae.* The lockkeeper glared in that direction, but said nothing. He and old John moved swiftly up front and Abby and Mike watched them use iron levers to open wickets or paddlegates that were underwater, near the bottom of the closed wooden gates. Released from the upper level, water flowed rapidly into the lower part of the lock chamber.

A rumble sounded from beneath the *Daisy Mae.* White bubbles foamed, churned, and gurgled up and 'round the sides of the boat. It rode slowly up the stone walls "almost as tight as a cork in a bottle," chuckled old John.

The deck lurched and rolled a bit underfoot. Thrown off balance, Abby and Mike clutched at Mrs. Clark who stood beside them. But she smiled at them and they grinned back, calmed.

The boat rose, like an elevator, up the sides of the lock. Stone by stone. The faces of the people on the locksides grew nearer and larger. Soon, the deck was level with the top of the lock.

"Come on!" shrilled Abby, waving to Mike, as she ran to the stern. They looked back, over the gate, and down at the water surface, about 8 feet below, where they had floated only minutes before. Each lock along the canal was a "water step" rising higher and higher as boats traveled up to Cumberland. On the return trip to sea level at Georgetown, the *Daisy Mae* would sink about 8 feet in each lock chamber, as the water was made to escape to lower levels through the gate wickets.

Captain Clark gripped the tiller of the *Daisy Mae* while the lockkeeper opened the forward gates wide, so the barge could be towed out of the lock.

Abby and Mike saw the canal curving clear and far ahead, with Bozo already straining his neck into his collar as he plodded up the towpath, pulling on the long rope. Led by old John, Bozo soon had the 125 feet of towrope taut and drawing the boat.

On the berm side, opposite the towpath, blocks of granite formed a wall separating Canal Road from the waterway. Captain Clark indicated a hill to the right of the barge.

"See that house on the hilltop, Mike?"

"Yes sir — the red brick one?"

"Correct. Well, that was once the home of Francis Scott Key. He was a lawyer. In 1814, during the War of 1812, he tried to release a physician-friend imprisoned by the British on one of their warships in the Patapsco River."

Captain Clark paused. He had noted that Mike bristled when the British were mentioned as enemies.

"Don't get your English back up, son," he said. "You're part American, too, you know," he chuckled, " . . . the good part."

Just then — almost beside the *Daisy Mae* — sounded the third blast from the tin horn of the impatient steersman.

Old John glanced back, then quickly unhitched the towing line from Bozo's harness. He lowered the towrope, tying its end to weeds low on the canal bank. Then he and Bozo stood out of the way, while the hurrying hoggee and his mule passed safely over the *Daisy Mae's* line and drew their boat up the canal.

Captain Clark's lips thinned as he waited for old John to hitch up Bozo and resume towing.

"Anyway," he continued, turning to Mike, "Key carried a flag of truce and a note from President Madison, asking that the doctor be freed. But the British 'kidnapped' Key and sailed upriver to bombard Fort McHenry in Baltimore Harbor. That's where Key wrote our national anthem, 'The Star-Spangled Banner' on the back of an envelope."

Captain Clark glanced sideways at Mike. "It may make you feel better, old chap (he emphasized the last two words, because they sounded so British), "to learn that Key's verses were later set to the tune of an old English drinking song called 'Anacreon in Heaven.' "

"I wonder if they still sing it in our pubs," mused Mike.

"Something for you — I mean, your father — to check up on, when you return home. Anyway, Key had moved into that red brick house on the hill in 1805. At least nine of his children were born in it. But they all left in the mid-1830s, because of all the activities and hubbub along the canal. Besides, the canal builders had destroyed Key's hanging gardens and the fruit trees growing on terraces that extended down to the Potomac. Nonetheless, Key remained a busy lawyer. He handled some affairs of the C & O Canal and took part in all sorts of social causes and civic events."

The *Daisy Mae* was now opposite the Aqueduct Bridge. Atop this structure there was once a great, water-filled trough made of wood; it carried towboats across the "River of Swans" (as the Piscataway Indians called it, long before white men came to know it as the Potowmack), down to Alexandria, Virginia.

Abby read aloud the quaint names, painted in red or gold letters on the sterns of boats waiting, lined up, to make the river crossing:

" . . . *Rough and Ready, Jenny Lind, Lady Lightfoot, Scow Lion, Fair Play, Good Intent* . . . and look!," she hooted, "those last two are named *Whirlwind* and *Lightning*. They haven't moved an inch since I sighted them."

She and Mike watched some sailboats near the bridge's stone abutments. The wind puffed out their gaily colored sails of red, green, yellow, and blue. Canoes glided along there, too, and Mike saw the paddles dip into the water, causing wavelets to flash and sparkle in the sunlight.

The *Daisy Mae* approached a rowboat anchored near the berm, or far, bank of the canal. In it sat a fisherman, holding a fishing rod. He was dozing off and his head jerked back, then dropped quickly to his chest.

"I say!" shouted Mike. "He's caught a fish." And Abby saw the line grow taut, then jiggle rapidly up and down in the water as the hooked fish tried to escape. The fisherman awoke suddenly. Apparently, he forgot where he was for the moment. He leaped quickly to his feet. The boat tipped on its side, the man lost his balance and went flying, with arms and legs outstretched, into the canal. He rose, spouting a stream of water from his mouth. Also some language the children could not understand, though Mrs. Clark clapped her hands over her ears, shut her eyes, and winced.

Her husband chuckled, with a wink at Mike: "Looks like the fish caught the man, doesn't it?"

They watched the fisherman scramble back into his boat and row rapidly to the towpath, in a hurry to get home, no doubt, and out of his wet clothes.

CHAPTER 2

Life on a Canal Boat.
Captain John Smith and Pocahontas.
Past Glories.

Canal families always had chores to perform and the crew of the *Daisy Mae*, in that summer of 1908, was no exception.

Abby was to help her mother with the meals, setting the table, washing the dishes and clothes, and other household duties. Mike, as a part-time hoggee, was to help drive and take care of Bozo and Lazy Bones. He was also to fetch the mules their food and water when old John was busy elsewhere.

"Those 'hay-burners,' " Old John had explained to Mike, "are half-horse, half-donkey. They're mighty powerful and need less care and food than horses. Oats, corn, and hay suits them just fine. And, of course, that pail of water apiece you'll be fetching them. A few hours of rest, then a mule's fresh and ready to hit the towpath for another stretch."

Mike's canal barge-home for the summer was 92 feet long. He figured he could stride from end to end, taking 40 "giant steps." The boat was 14½ feet wide and drew almost 5 feet of water when fully loaded with cargo.

Bozo and Lazy Bones shared the stable-shelter on the front deck. The "hayhouse" or bunkhouse, was in the center of the boat. The 14 hatches covering the hold could be removed, revealing a great cavity, where 120 tons (that came to 240,000 pounds!) of goods could be stored.

The family quarters, under a canvas, sunscreening, awning at the stern, looked like a tiny house from the outside. A tin smoke chimney stuck up from the rooftop. Gingham curtains were hung in the glazed windows. Slatted, wooden shutters were pinned open, but could be closed for privacy. The interior of the cabin measured only 10 by 14 feet — about the size of a bedroom in a house on land.

In a recess of the cabin enclosure, three steps led below the deck, first to a galley, or kitchen, where a black iron cookstove stood. It burned corncobs when a quick, hot, fire was needed to prepare a meal, and coal when a slow, long-burning, fire was required.

Next to the stove stood a small table and four chairs. The tabletop was hinged to provide more room, when it was not in use. One of the table's legs could be swung out to prop the tabletop when meals were served. Pegs for hanging clothes were driven into the walls between the windows.

Near the stove, a door led into a tiny "stateroom" where Captain and Mrs. Clark slept. Their bed was so narrow that if Mrs. Clark wanted to get out in the morning to prepare breakfast, she would have to warn her half-asleep husband: "Lay still now; I need to get out" — then crawl over him.

Abby slept in an outside bunk of the cabin. And old John and Mike slept in the hayhouse, in hammocks slung from hooks on the walls. Passengers, if any, also bunked there — though most preferred to sleep on blankets spread on deck in good weather.

A china cupboard and storage spaces were built into wall compartments of the cabin. A blue-checkered oilcloth, smelling heavily of linseed oil, covered the table. And in its center, at the moment, stood a glass jar of water. In it, Abby had stuck a sweet potato, and, already, white roots crowded the bottom, and green shoots and leaves studded the brown plant. Each side wall of the cabin had a bracket that held a kerosene lamp topped by a glass chimney. Not an inch of space was wasted in the cabin. At night, a curtain could be pulled across the cabin, dividing it into two compartments, for privacy.

Captain Clark always rested one mule aboard the *Daisy Mae*, while the other towed the barge. Sometimes, old John would take a nap on the hay heaped next to the mule stall. He would talk himself to sleep, telling Bozo or Lazy Bones all sorts of adventures — real or imagined — that he had had as a boy along the canal. And whichever mule rested there at the time would open its mouth wide, bare its teeth and bray and shake its head up and down, as though it understood every word. But the mule really wished old John would be quiet and let it rest, for it, too, was exhausted from dragging the heavy boat.

Indeed, while most boats traveled "light" up-canal, hauling grocery supplies for general stores along the river and canal route, the hold of the *Daisy Mae* was filled on this trip with all sorts of merchandise: Furniture, machinery, stoves, coal scuttles, shoes, crates of hardware, plows, grindstones, bolts of cloth, medicines — things needed by storekeepers, farmers, and townspeople upstream and especially in the Cumberland area who weren't in any all-fired hurry for deliveries.

The canal only went as far as Cumberland in the Allegheny Mountains. Had it terminated at Pittsburgh, as originally planned, perhaps some railroad trains there might have chug-chugged up and carried away the cargo and some leisurely travelers — westward across the vast plains and beyond.

"Some of the canal cargoes and passengers," explained Captain Clark to Mike, "then might have ended up beyond the Rocky Mountains, in San Francisco, on the banks of the blue Pacific Ocean, thousands of miles away from Georgetown."

On the return trip from Cumberland, the *Daisy Mae*, as usual, would carry wheat, corn, lumber, and coal (known as "Black Gold" to the boatmen). Almost all canal boats hauled coal, but Mrs. Clark often talked her husband out of it. Although the pay was better, the black dust always sifted into everything: clothes, bunks, dishes — anything that was not carefully covered, and many things that were.

"Besides," she argued, "the coal dust is bad for your lungs."

Sometimes the *Daisy Mae* would transport a passenger or two. Such travelers were good company, said the captain, for they were in no hurry and often carried a musical instrument, like a fiddle, or a banjo, a harmonica, a Jew's-harp, a fife, or a flute. Or they would hum or toot a tune through a pocketcomb covered with thin tissue paper. Anything to while away the long hours of the night and the waiting times at busy locks. Besides, the *Daisy Mae* moved at a top speed of only 3 miles an hour.

A mule usually towed the boat for about 6 hours, then rode aboard for 6 hours, while the spare mule took over the pulling job. Sometimes, when approaching a lock, old John would give the towing mule a rest. He would unhitch it, pat its flanks, and talk gently to it. Then he would grab the bowline and, helped by the lockkeeper, would loop the rope around a wooden snubbing post. This would slow up the boat and control it so that its fenders wouldn't damage the lock walls or the wooden gates. The post also made it easier to guide the boat into the narrow lock entrances.

Miles from the cobblestoned streets of Georgetown, Captain Clark glanced up from watching Abby and Mike play "Lotto" on the deck. (This was a game like present-day "Bingo"). The captain noted that old John, leading Lazy Bones, had absent-mindedly neglected to warn Lock 6 of their approach. Letting go of the tiller, he stretched himself to full height and cupped his hands around his mouth.

"Hey-ey,ey, lock!" he yelled. "Hey-ey-ey, lock! Lock ready! Lock ready!"

And far up the canal, the lockkeeper heard the cry and came bustling out of his house toward the gates. Soon the *Daisy Mae* was being locked through the sixth "water step." This was near the first of six dams; each one fed water into the canal along its length of 185 miles.

Mike had watched the entire locking through operation. Now he stood with his hand on the tiller, next to his uncle, "helping" him steer.

"What's the history of this place, uncle?" Mike wanted to know. "Didn't redskins live around here?"

Captain Clark beamed. "Indeed they did! The Algonquin Indians, headed by Chief Powhatan, had their fishing villages all around here. But you should know, Michael, that your countryman, Captain John Smith explored this area, over 300 years ago. According to most legends, he fell into the hands of Powhatan and was about to be killed, when the chief's daughter, Princess Pocahontas, saved Smith's life. The Indian maiden — or redskin as you would call her — later married another Englishman, John Rolfe. Rolfe took her home to England and she lived in London as an 'American princess.' She died in your native land, Mike, but many prominent Americans are descended from her.

"Anyway, Captain Smith claimed that he boated up the Potomac as far as he could go — until the many rocks in the river stopped his boat just about there — over to the left. That's the place we now call Little Falls." He pointed upriver in the direction of the first dam's north abutment. "That location was also the birthplace of the Chesapeake and Ohio Canal. In 1828."

Mike did some rapid mental arithmetic. "That was exactly 80 years ago," he said.

"Exactly," said his uncle, slapping the tiller in approval. "John Quincy Adams was our 6th President at the time. Tamping down a high silk hat over his head, he boarded a steamboat tied up in the swampy flats near the White House at Lock Cove. The noisy little vessel puff-chugged its way upriver a short distance. Then Adams came ashore and walked over to a boat on the old Potowmack Canal built by George Washington. The Presidential party was transported above what is now Lock 6, almost 6 miles west of Washington City.

"President Adams made a flowery speech," said the captain, waving his arms grandly. "He proclaimed that a canal would be built to stretch from Georgetown to the headwaters of the Ohio River — within 3 years. (It actually took 22 years to get through to Cumberland, Maryland, and it never reached Pittsburgh). This grand waterway, promised the President, would be more famous than all the wonders of the ancient world. He mentioned the Pyramids of Egypt, the Colossus of Rhodes, even the Great Wall of China."

Mike whistled. "Wasn't he exaggerating?"

"He sure was laying it on a bit thick," agreed Captain Clark. "But being a good politician, Adams only built up the fact that the canal was to be another engineering marvel of the times — which it really became, you know.

"Anyway, the crowd cheered the President's long-winded speech: 'Hip-hip, hooray; hip-hip, hooray; hip-hip, hooray.' They also shot off fireworks in celebration of the Fourth of July — our Independence Day from your King George III. (Mike *did* bat an eye — but he was getting used to the teasing of his uncle.)

"Then," continued Captain Clark, "John Quincy did a practical thing. He grabbed a spade and drove it into the ground. But it glanced off a tree root! Again, he thrust the spade into the earth and stamped down on it with the heel of his boot. But the root was tough — it wouldn't give in. Not even to a President of the United States. So Adams removed his greatcoat, spat on his hands and, this time, he cut through the root and turned the first shovelful of dirt.

"That small hole in the soil became the canal. With the labor of thousands of immigrant Irish and German workmen. They used axes, stump-pullers, and horse- and ox-drawn scoops and scrapers. Those immigrants faced up bravely to sickness, disease, and brutal work. Even death. Many of them died of cholera and malaria. Or riots, when they weren't paid. Many were killed in drunken brawls, after they were paid. And the mounds of their unmarked graves dotted the canal route. But the canal was built — and built well by the new Americans."

"Thank you, uncle, for that bit of history," said Mike, now properly impressed. "And they dug it all without steam shovels?"

"Yes, but they did use black gunpowder to blast the rock out of their way. It was all hard and dangerous work."

"Much harder, I'm sure," agreed Mike, pleasantly, "than gliding along on the canal as we're now doing."

He left his uncle to go up to the bow where Abby was watching Lazy Bones amble over the towpath. The towrope stretched from the mule's harness and dipped into the canal or slithered over the weeds and flowers along the edges. Sometimes the rope tore off a hibiscus plant and it floated away, past the barge, making Abby feel sad because it was so pretty and was destroyed. But there were patches of other flowers up ahead: Black-eyed Susans, daisies, Dutchman's-breeches, violets, trout lilies, blue phlox, mint, and a number of others. Abby grew happy again, dazzled by the sight.

Mike came up from behind her and tapped her shoulder.

"Look at that turtle," he said. It was sunning itself on a submerged log near the berm. Awakened from its nap, it plopped into the water as the barge drew near.

"I say, Abby!" cried Mike. "There goes an eel." They both watched it wriggle away underwater, like a snake. Then they spotted a dragonfly hovering over the head of a bullfrog that seemed asleep in a shallow, rocky pool near the shore. But the frog moved swiftly out of a shadow and flicked its long tongue. The dragonfly disappeared down the throat of the frog.

"Ca-woork!" it went and closed its eyes.

The boat passed rolling pastures where cows and sheep grazed under tall shade trees. Some quail, startled by Lazy Bones on the towpath, shot into the air, then settled back to earth, under an elderberry bush.

The warm sunlight filtered through the trees that bordered the canal and made golden patches on the dark-green water. It looked cool and inviting, especially since the humid heat made them all feel drowsy.

Mrs. Clark had promised Abby and Mike they could go swimming in the canal, behind and near the boat's stern, where she could watch them "after we've passed all the warehouses, markets, taverns, and houses on the outskirts of the city, and reached the open meadows of the countryside."

Now she came on deck from the stuffy cabin and mopped the perspiration off her forehead with her apron.

"My stars, it's hot and sticky," she sighed to herself, then she called: "Children! You may swim for awhile."

"Yea!" they yelled. Since they wore only rough old clothes and were already barefoot, they simply dove headfirst into the water and came up to the surface to paddle about, cool and refreshed. Mike floated on his back and breathed a long "Ah-a-a-a-a . . ."

"Come on, Mike!" challenged Abby. "Let's race up to old John on the towpath."

They churned through the water, spurted past the slow-moving boat, ducked under the towrope, and bobbed up near old John. He was singing softly to himself and leading Lazy Bones by the bridle.

"Hey, there!" he cried at their sudden appearance. He pretended to be startled. "You had me scared into popeyes. Couldn't figure how two big fish like you got in this canal."

Abby and Mike laughed. They splashed water onto the towpath and over Lazy Bones. The mule brayed with pleasure as the water struck its back and dribbled down its sides. Abby and Mike floated around, spouting water like baby whales.

Abby sang "School Days," a popular new song. She had learned it from a graphophone, or talking machine, in a music store in Georgetown.

Mike clapped his hands when Abby finished. But he also splashed wavelets over her face and into her open mouth and eyes. Abby coughed, gasped, and splashed water back at her cousin.

The *Daisy Mae* caught up with them and passed by.

"Abby! Mike!" called Mrs. Clark from the stern. "Keep up with the boat." She looked a bit anxious. But Captain Clark stood calmly by the tiller, guiding the barge and smoking his pipe, contentedly. Little waves rippled past the bow's stempost, around the fenders, and curled along the hull of the *Daisy Mae*.

The youngsters swam easily behind the rudder, shouting, waving, calling to Mrs. Clark. She sat on a low stool next to the cabin. On the deck at her feet were two buckets. One was filled with water and dirt-covered potatoes, the other with clean water for the peeled ones. Occasionally, she would lift her head to check on the swimmers, while preparing the evening meal. "Keep close to the boat," she cautioned, beckoning to them every few minutes with the paring knife in her hand.

Lazy Bones plodded up the towpath and pumped his head up and down, up and down. His flanks moved smoothly and the towrope, hanging from his collar, dipped and splashed and rose and fell in the water, especially when the boat wound around the bends in the canal.

After awhile, the weight of their waterlogged "bathing suits" made Abby and Mike tired.

"Please, father," called Abby, lifting one arm high. Captain Clark bent over and helped his daughter, then Mike, climb over the boat's water-rail.

"Thank you," they sighed, gratefully, and stretched out quietly on the hatches drying and sunning themselves.

The barge passed near Sycamore Island, off to the left.

"What's that, uncle?" asked Mike, raising himself on one elbow and pointing at the rows on rows of canoes, bottoms up, in the sheds beside a clubhouse.

"That's the famous Sycamore Club," said the captain. "For canoeists and sportsmen. He leaned his weight against the tiller and the boat moved closer to the towpath. "Near that club," he said, "once stood strange towers, minarets, galleries, and outlooks. They were made of 40,000 red cedar logs. Stood on that hillside — there — between the canal and Conduit Road, overlooking Sycamore Island and the Potomac."

"But I see nothing there now," said Mike.

"Right," said his uncle. "Nothing but young trees and undergrowth. However, elegant carriages once did pull up there to the rustic entrance of the Pa-Tow-Wo-Meck Cafe. Diners came to Glen Echo Heights all the way from Washington, until 1890. That's when the whole fantastic collection of buildings burned down."

"Did you ever take mother to dinner there?" Abby wanted to know.

"No," said her father. "But I wish I had; then I could describe the place from personal experience. I understand that it was a quiet, relaxing spot with breathtaking scenery everywhere one looked." He paused to puff awhile on his pipe. Then he continued:

"Shortly after the Pa-Tow-Wo-Meck Cafe and all the other odd structures burned down, a Chautauqua, or National Culture Center of the United States was planned nearby. In a picturesque town called Glen Echo. Culture-seekers settled there, building cottages along tree-shaded avenues. The roads and streets were named for noted universities and colleges — like Yale, Harvard, Wellesley — and even — Mike, your famous English Oxford University.

"A great auditorium was built of stone, to attract America's most famous preachers, lecturers, authors, and entertainers. Eight thousand people could be thrilled at one time by the performances or by music from a giant organ. Then came trouble . . ."

By this time, the captain was excited and was waving his arms, forgetting he was supposed to be guiding the boat.

"For pity sakes!" cried Mrs. Clark, suddenly bobbing her head up from the rear cabin. "You almost scraped us all off against the bank that time." Then she laughed. "Perhaps I'd better steer the boat while you lecture the poor children."

"No, dear, thank you," said the captain. "Just mind your potatoes. I'll steer and talk. And if I add excitement to my lectures — why that'll keep my 'students' from dozing off. Right, kids?"

"Righto!" cried Mike.

"Well, *I'm* not so sure," said Abby.

Her father pretended to look pained. "These modern children," he grumbled. "No appreciation . . ."

"Oh, please continue, Uncle," urged Mike. "You were at an exciting part. What happened to the Chau-took . . . ah, Chaustalk . . . uh, the place you said?"

"Ah!" said the captain. "There's a smart young fellow who appreciates me. Well, sir — water from the Minnehaha Creek ran under the huge Chautauqua auditorium to power its organ. But mosquitoes bred in the stone grottoes beneath the amphitheatre. And the bites of those pesky little female beasties carried malaria bacteria. So many people got sick that attendance slumped — like that!" He gestured his hand into a steep dive.

"And the magnificent hotel of stone that was to be built, like a castle, high on a hillside above the homes of wealthy and important people . . . well, that was never constructed.

"Ex-President Grover Cleveland and other government officials had bought land and planned to live in the community. But the dream of a national cultural center was never realized. The stone buildings and the auditorium of the Chautauqua were converted into an amusement park. "We'll go there tomorrow, after we visit my old friend, Clara Barton."

After passing through Lock 7, Captain Clark tied up the barge for the night, in a jungle of fragrant honeysuckle and towering sycamore trees covered with English ivy. It was on the berm side of the canal, near the dock of the American Red Cross Headquarters. Like many wide places throughout the length of the waterway, it was safely out of the way of hurrying barges that transported cargoes without stopping.

Old John led Lazy Bones over the mule fall-board — a sort of gangplank — and into the stable. Mike fed both mules and filled their pails with fresh water. Then he tidied up the stable, washed himself thoroughly, and joined the Clarks at their evening meal.

Afterward, everyone rested on deck, enjoying the twilight calm. The sun set and the dim stars grew larger and brighter in the sky. A woodthrush poured out its heart in a song that soared into the air then fell glittering over the woodlands skirting the canal. The haunting notes of a whippoorwill: "Whip . . . poor . . . will! Whip . . . poor . . . will!" rose and fell on the quietness. And every once in awhile a firefly winked its lanternlight.

Abby and Mike got up to watch the hoggees on the towpath, going past the *Daisy Mae.* They held the mule harnesses in one hand, lighted lanterns in the other. They trudged by, whistling, into the darkness, followed by the lights of their boats wavering over the water.

Soon, Abby and Mike tired of swatting at the gnats and mosquitoes that swarmed and buzzed around their heads and through their hair and into their ears. They went into the cabin. Old John was dozing in a chair, Captain Clark was reading *The Washington Post* by a kerosene lamp resting on the table, next to Abby's flowering sweet potato plant. Mrs. Clark was busily darning socks. She looked up and smiled at the children.

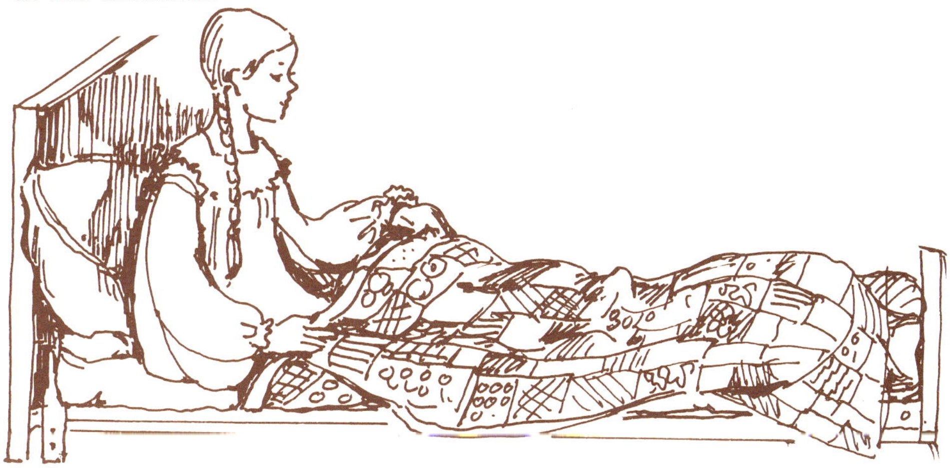

"You both look sleepy. Ready for bed?"

They nodded. Mrs. Clark helped hang their clothes on pegs and watched Abby climb wearily into her outside bunk. She kissed Abby "Good night" and drew the curtain down the center of the cabin. Then, carrying a lantern, she went to the bunkhouse with Mike and held the hammock still for him while he climbed into it.

"Good night; pleasant dreams," she said, smiling, and was gone.

Mike swayed back and forth in the dark, listening to the sounds of the night: the warbling of nested birds; the ca-wo-r-k! ca-wo-r-k! of frogs; the who-O-O of an owl, and the gentle lap-lap of water against the hull of the *Daisy Mae.*

He fell into a deep sleep . . .

AMERICAN
RED CROSS

CHAPTER 3

We Meet Clara Barton.
Glen Echo Amusement Park. Union Arch.
Cabin John Hotel.

The morning sun streamed through the cabin window and a warm ray tickled Abby's nose. She awoke with a loud sneeze: "Ah-choo!"

"Bless you!" Her mother glanced up from the skirt she was ironing (using the table as a "board," and a towel as padding). "We must have two Lazy Bones: you and the mule. Everyone else has had breakfast and gone up to the Red Cross dock. They're unloading supplies. Miss Barton is there right now, checking off the boxes. See her chatting with your father and Michael?"

"Yes, ma'am," said Abby, gulping down a mouthful of warm oatmeal. "The small lady, dressed in dark clothes . . . Why does father call her the 'Angel of the Battlefield'?"

"Oh, he often calls her the 'little living saint.' But she was indeed an 'Angel of the Battlefield' during the War Between the States. She treated both the wounded Union soldiers and captured Confederates with the same gentle skill and care.

"Miss Barton is quite old now, but you'll still find her spry and perky. Scoot up there, now, and pay your respects, before she takes off on a relief mission 'cross country or 'round the world. You know, it's happened every time you've passed here before on your vacations from school."

Mrs. Clark gave Abby a handful of raisins to nibble and struck her bottom lightly with an open palm. "Scoot!" she said. "I've work to do."

Abby swung over the side of the barge and leaped onto the dock. Her father put his arm around her shoulder and turned her to face Miss Barton. She was a bit taller than Michael, but shorter than Abby. A large, enameled, Red Cross emblem was pinned in the collar at her throat. She had been watching workmen unload boxes from a barge, while nurses dressed in white uniforms with red cloth crosses on the front of their headdresses and left armbands, checked off items on long sheets of paper with their pencils.

The little lady's bright eyes sparkled as she smiled and held out her hand. "We meet at last! How do you do, Miss Abigail? I am indeed happy to see you."

Abby curtsied, then rested her hand in that of the "Angel of the Battlefield." Abby noted that Miss Barton's keen, deepset, eyes and stern face had a way of suddenly turning smiley and kind, just like her mother's. "I do nicely, thank you," she replied. "But you don't look so old; your hair isn't white."

Miss Barton threw back her head and laughed. The corners of her eyes crinkled. "Out of the mouths of babes! No, my hair's not white; nonetheless, I'm twice as old as your father. Come now with me, for I promised him I'd show you and your English cousin, Michael, through my house. Later, we can go to the Glen Echo Amusement Park; when it opens this afternoon."

Then followed an unforgettable day for the children. Miss Barton was a small bundle of energy. She led Captain Clark and the children up the hill to her great house of 36 rooms. She showed off the various shrubs, plants, and flowers she had planted around the house. The pungent smell of boxwood bushes hung in the air.

Miss Barton lifted her skirt daintily, climbed the front stairs, and led her guests into a long, crowded, hall. Here, groups of ladies in nurses' uniforms were rolling bandages, cutting gauze, and working at sewing machines and typewriters.

"Looks like a Mississippi steamboat, doesn't it?" Miss Barton waved her hand at the open, railed, galleries overhead. "Especially with those three decks and the captain's quarters and wheelhouse up above."

"Gosh, it sure does!" exclaimed Abby, bending her head backward.

"Quite so," agreed Mike, also staring up at the "passengers' staterooms." These, explained Miss Barton, provided living quarters and offices for her Red Cross staff.

Mike stepped back to get a better view of the suspended ceiling of the "wheelhouse." He bumped into a Red Cross nurse at the bandages table. The roll of gauze she held in her hand went sailing across the floor, unspooling into a long white ribbon.

The lady glared at Mike. "Clumsy!" she muttered. Then her expression changed quickly to a toothy grin, for she had caught Miss Barton's eye.

Abby jabbed Mike with her elbow. "Beg the nurse's pardon, you silly," she whispered.

The youth did so, adding a deep bow. This gesture immediately softened the nurse's manner and drew chuckles from Captain Clark and Miss Barton.

"What beautiful woodwork," remarked the captain. He ran his hand over the closet walls and stooped to examine the hemlock flooring. "I can't help but always admire it."

"Yes," said Miss Barton, her face lighting up with enthusiasm, "it deserves your admiration. You see, Michael and Abigail," she turned to her young visitors, "much of the lumber and flooring in this building came down the canal on barges like your *Daisy Mae*, after the Johnstown Flood of 1889. The Pennsylvania refugees of that disaster had to live in temporary wood barracks. Later, they dismantled the structures and sent some of the lumber here to Red Cross Headquarters as a gift — because of our relief work among them."

Stepping to the walls, Miss Barton opened some panels. The walls were false! They were really deep closets, crammed with first aid supplies, clothing, nails, shovels, picks, blankets, and preserves and canned goods of all kinds — all ready to send anywhere in the world — whenever and wherever the call came for help.

From a shelf, Miss Barton took an interesting container made of woven cane. She unlatched the brass clasp on the lid. Inside, in three recesses covered with red velour, nestled a china teapot and two cups with covers.

"This was my treasure and comfort on the battlefields of the Civil War," explained the famous lady, patting the chest.

"Please. May I handle the pot and cups?" Abby asked, surprising her father, because her usual tomboyish manner was absent.

"Of course. Of course," smiled Miss Barton, delighted that Abby was so interested.

The girl brushed some stray hairs out of her eyes. She removed the chinaware pieces and turned them this way and that. She examined the Oriental designs of long-tailed purple birds and pink chrysanthemums. She wondered what the generals and their soldiers thought when they saw the chinaware used on the battlefields.

Abby stroked the velour covering inside the case; it felt like the smooth fur of a kitten. Then she carefully returned the china and closed the lid.

"That's a very beautiful china tea set, Miss Barton," she said. "Thanks for showing it to us."

Abby and Mike were impressed, too, by the letters the Civil War heroine had received from President Abraham Lincoln and many famous generals and Congressmen. They admired a goldleaf table with an onyx top from the French Government. Also a handsome gold chair with a seat of brocade cloth.

"My good friend, the Grand Duchess Louise of Baden, gave me this chair," said Miss Barton. "She was a German princess."

Mike's dark eyes widened. He whistled. "You mean a *real* princess, like the ones we have in England . . . ?"

"Like in fairy tales?" Abby interrupted.

Miss Barton chuckled. "Oh, my friend was real enough, and so is this gift of hers. Here, sit on it, while I show you some of the medals and keepsakes I've received from foreign kings and governments." She smiled up at Captain Clark. "I know you have seen these before," she said, apologetically. "But children do like to examine pretty ribboned things of gold, silver, and bronze."

They saw, handled, and admired other objects and curios: the oil paintings made by Miss Barton, her field trunk that opened into a cot, the ledgers filled with her neatly penned lists of missing Civil War soldiers, which President Lincoln had asked Miss Barton to prepare; also the many strange and exotic souvenirs collected on her visits to distant lands and on her lecture tours in America.

A visit to the kitchen produced a handful of cookies for each youngster. Also a glass of cold milk from the pitcher that stood on a cake of ice inside the top of the icebox. From Miss Barton's sunny bedroom, they could see the purple-blossomed magnolia trees beside the house and, beyond it, the amusement park nearby.

Already, they could hear the laughter of children riding the carousel, and the sound of music from the calliope.

"Come!" said Miss Barton, shouldering a white silk parasol. And the little lady was off, leading the party toward the amusement park.

Abby and Mike immediately jumped aboard the carousel, or merry-go-round. They were fascinated by the gaily painted wooden horses and zebras, the ostriches, lions, tigers, and rabbits. And what clown faces and mirrors covered the glittering top of the carousel! Captain Clark and Miss Barton sat sedately on one of the benches shaped like a chariot, while the great wheel turned to the tunes from the band organ.

The youngsters gripped the brass poles and rode the charging wooden animals up and down; up and down. They were whirled 'round and 'round until they almost grew dizzy. The pounding on the drum and the zing! clash! of the brass cymbals on the big calliope matched their excitement. Especially when Mike leaned out from astride his handsome "galloping" pony, so unlike either Bozo or Lazy Bones. He grabbed for the shining brass ring which would win him a free ride.

Mike stretched too far and went sliding off the pony's back. His shirt ripped open as he fell flat on the floor between an ostrich and a lion. He got up on his knees. "Ouch!" he cried, bumping his head against the hoof of a rearing horse.

Red-faced, embarrassed, he wiggled to his feet in front of Miss Barton. She promptly reached into her large handbag, found a safety pin, and closed Mike's torn shirt.

Captain Clark shook his head. "Mmm! Mmm!" he grunted. "How can one boy get into so many scrapes? And speaking of scrapes — I see you've skinned your left elbow."

Miss Barton reached again into her handbag for a small first-aid kit. "Always carry this," she said. "Now, young man, this will smart a bit." And she applied some iodine. "There you are, Michael, good as new."

"Thank you very much indeed, Ma'am," he said, politely, though his skinned elbow felt on fire.

"Now we can call you the 'Angel of the Merry-Go-Round,' as well as 'Angel of the Battlefield,' can't we?" said Abby.

Miss Barton threw back her head and laughed. "If you wish," she said.

Mike climbed back on his pony and leaned outward. He gritted his teeth. At the next revolution of the carousel, he stretched out his right hand — and the "free ride" brass ring appeared to leap into his fingers.

"By golly!" exclaimed Captain Clark. "He's got it." He pretended he didn't like to pay everyone else's fare around again — but he did.

He even bought four paper cones of pink cotton candy. And they all laughed when some of it stuck around Mike's mouth and chin and Abby stared at him, saying: "You look like a little old man with a pink beard and mustache."

When they parted at the entrance to her Red Cross Headquarters-home, Miss Barton patted Abby on her shoulder and gave her a lovely little Red Cross nurse's pin. In its center was an enameled red cross. It was identical to the one Clara Barton wore at her throat collar. She gave Mike a similar button, to be fitted into the lapel hole of his best suit coat.

Captain Clark towered over Miss Barton; she looked quite frail and small in her black dress. Suddenly, he took her thin old hand in his and, bending low, kissed it, gallantly. They smiled a silent farewell at each other.

"Goodby, children," waved Clara Barton, as they skipped down the hill to the loading dock. "We've had such great fun together; you must come back soon for more."

Mrs. Clark listened to the adventures of Abby and Mike. She admired the pin and button given them by the grand old lady. When the youngsters stopped for breath, Mrs. Clark was able to say: "Old John has Bozo fed and ready on the towpath. As for me — your excitement has also tired me; I must take a nap before I prepare dinner."

It was late afternoon when the *Daisy Mae* moved away from the Glen Echo dock and was steered toward Lock 8, below the town of Cabin John.

But several barges were stopped ahead. Captain Clark spotted a produce stall ashore. "Halloo, John!" he called, "Take care of the boat and mules. We'll be back soon."

Some men on the berm threw them a line and helped swing the stern gently against the shore, where the boat was secured to an old tree trunk. The captain helped his wife and the youngsters over the aft rail and onto a tiny dock next to the produce stand. It displayed for sale smoked hams, vegetables, fruits, and jars of food that the lock-keeper's wife had grown in her garden, or had preserved.

The stone lockhouse was whitewashed, so it would be easier to see at night. It had a sloping cellar door outside, with a frame summer kitchen added onto one side of the building. A porch extended along another wall. Small outbuildings housed hogs and chickens (which provided ham, pork, sausages, and eggs), or protected axes, plows, spades, and rakes from the weather. Two cows gave fresh milk, from which were made cream, butter, and farmer's cheese. And, sometimes, even fresh ice cream was mixed, frozen, and sold to delighted travelers or canallers.

A gray horse, munching on blades of grass nearby, provided transportation. A small, rock-bordered pond "jumped" with its stock of fish, frogs, and ducks. Even eels. So this lockkeeper's family was quite comfortable in all respects. Particularly since their lockhouse was built of stone — not logs or wood frame, which could get mighty cold in wintertime.

"Come inside," invited the lockkeeper's wife, leaving her grown daughter to tend the stand and sell the produce to the Clarks. The woman had watched Mike stroll around the lockhouse, peering curiously into a window, and at the horseshoe nailed over the doorway "to bring good luck." She beckoned to Abby. "You can come along, too, dearie," she said.

The youngsters strolled through the tidy rooms and admired the large stone fireplace that could heat both the living room and the kitchen. The two family bedrooms were upstairs — small but snug — and were reached by a narrow stairway. Each contained a bed, a chest of drawers, and a mirror over a stand that held a white water pitcher and a bowl and a white commode or "slop jar."

The three then went down into the cellar. Here, shelves against the walls were loaded with glass jars of preserved pears, apples, tomatoes, beets, pickles, stringbeans, corn, squash, and peas — all grown in the garden patch near the canal. Also jars of blackberry, strawberry, and huckleberry preserves.

The lockkeeper's wife glanced at Mike's mouth working. She chuckled, "Goodness me! You're licking your chops, son."

Laden with baskets of the goodies, the family waddled back to the *Daisy Mae*. Mrs. Clark, however, was muttering to herself: "Fifty cents! How awful . . . 50c for a smoked ham. Land sakes alive! 'Tain't worth more that 40c. But it did look good — and I need it for our dinner tonight."

Back on the boat, Mrs. Clark and Abby stored the purchases in a quarterdeck closet in the stern, reached by a small door from the gallery. Then they began preparing dinner, while the captain manned the tiller. Mike and old John led the mule up the towpath. They ambled slowly,. for canal traffic was still heavy up ahead.

Even though there was no lock in sight, Mike was humming softly to himself a little ditty taught him by old John.

"Come on, Mike," urged the kindly black man, "you're the assistant hoggee. Don't just hum it — sing it! Better still — let's both sing it — loud and clear."

And they did:

"Lock ready! Lock ready!
Lock ready for me;
Get out of that bed
You sleepy head —
And get that lock ready for me!"

There was a loud clapping of hands behind them. Mike turned and waved to the "audience."

"I say, John," called Captain Clark. "Can you spare your singing young friend — I mean, Michael — for awhile? I'd like to tell him about some of the sights just ahead."

"Coming, uncle," said Mike. He patted Bozo's head. "Bye-bye, old chap," he said. "See you later, Mr. Manson. And thanks for teaching me that ditty."

The *Daisy Mae* was steered close to the bank and Mike scrambled aboard.

"Settle yourself comfortably on deck," said the captain. "Because soon you'll be seeing a fairy wonderland of colored lights, parks, and rustic buildings built around the famous Cabin John Hotel. But first — framing the beautiful picture ready for you — will be the Cabin John Bridge. A dam and a special lock were built here on the canal, so that great blocks of stone could be floated down the canal, then up Cabin John Creek, to construct the bridge. It was first known as the Union Arch; for many years it was the longest single-span stone arch in the world."

"Indeed, uncle? In the whole world?" Mike was impressed.

"Yessiree. And it's not only a bridge but an aqueduct."

"An 'aqueduct' — why, I say, that's what carries water. I saw some ruined aqueducts in England. The Romans built them."

"Good, Michael. You know your history. Now, then — our Nation's Capital — the City of Washington — gets its water supply from the Potomac River at Great Falls — we'll be there tomorrow. That water flows through a great conduit underground until it gets to the bridge . . .

"There it is, now. See those wagons moving across the top? Well, just below their wheels, within the floor of the bridge, is a huge iron pipe. And millions of gallons of water pour through it every day, headed for Washington."

" '. . . Millions of gallons of water . . . every day' " repeated Mike, his mouth slack with dismay. "All that? I say-y! Why, fetching the mules a bucket filled with only 5 gallons of water makes my hands smart and feel rough."

He stood up for a better look, marveling at the graceful span far up the creek. His uncle explained that soldiers had guarded the bridge during the Civil War and that the names of both the then President of the United States, Abraham Lincoln, and Jefferson Davis, who became President of the Confederate States of America — enemy leaders of the North and South — had once been chiseled together into a stone tablet set in the wall. Davis' name was cut out, during the Civil War, because of Northern hatred, said Captain Clark.

"I see the tablet. I see it!" cried Mike. "And the blank spot on it, too — there, on the left side of the bridge, behind the trees."

"Correct," said Captain Clark. "And now, son, there's the Cabin John Hotel and its grounds, framed by the bridge. Look at all those fancy buildings — all with that gingerbread architecture. And those colored electric lamps strung all over the iron bridge across the creek.

"Some view, hey? John Philip Sousa, the world-famous 'March King' and leader of the United States Marine Band often plays his tunes there. And his inspiring marches certainly make our spines tingle with pride in being Americans.

"The Hotel also has a marvelous orchestrion — that's a kind of music box that plays large records made of steel, with perforated holes that trigger the playing of various instruments. The orchestrion was imported from Europe and . . . good glory! . . . there it plays now."

They all listened while the captain steered to the right, out of the path of passing barges, then signaled to old John to unhitch the mule and rest awhile on the towpath.

The music-box sounds of Strauss' Blue Danube Waltz tinkled from the ornate rustic tower and floated over the valley. The notes echoed and re-echoed from the hillsides, from the vineyards on the creek banks, and settled softly over the canal and enveloped the listeners staring up at the romantic view.

Abby watched her parents grope for each other's hands. She, too, felt shivery, enchanted by the twilight. A mist shrouded the Cabin John Bridge in the background, and shadowy, far-off people moved through the garden mazes, and over a footbridge built of logs in the foreground, in time to the music. The spectacle looked like some kingdom in a Hans Christian Anderson fairy tale.

And Mike looked around him and recalled a visit to a magical theatre in London with his parents. He remembered the stage with its glittery scenery, the soft music playing, and all those excited faces in the audience. They now matched those around him at this real wonderland along the C & O Canal.

Captain Clark explained that Joseph Bobinger, a stonemason from Alsace-Lorraine in France and his German wife, Rosa, came to Cabin John before the Civil War. Joseph worked on the Union Arch aqueduct; Rosa fed the engineers, inspectors, and masons. The immigrants prospered, became American citizens and bought both sides of the Cabin John valley, except for the government-owned area around the bridge. They built their fabulous hotel on a hillside next to the bridge.

Rosa invented and served Maryland Fried Chicken, continued the captain. She served it on the finest Bohemian china, he said, with Cabin John-grown wines sparkling in ruby and clear crystal goblets. The handsome dining halls, gleaming silverware, and elegant service attracted most of the Nation's notables, when they came to Washington City. "And distinguished foreigners still come here," said the captain, nodding at Mike, "though Joseph and Rosa died years ago and their son, William, now operates the hotel and its grounds."

The captain stood shading his eyes and staring at a gazebo, or rustic summerhouse, some distance up a hillside near the canal. A big, stout man had left the gazebo and was strolling in a flower garden, directing the cutting of some blooms by a gardener.

"Yes; that's Will Bobinger — one of the hotel owners," said the captain, cupping his hands around his mouth. "Ahoy there, Will!" he cried.

The heavyset man turned and his face burst into a smile. He waved his arm and came down to the canal. Reaching across the water-rail, he shook the boatman's hand.

"Captain Clark! Delighted!" His walrus mustache and the heavy gold watch chain across his vest jiggled as he pumped the captain's hand. He looked very formal in his black silk suit, upright collar, and black bowtie topping a gleaming white shirtfront.

"Just like your father, I see," greeted Captain Clark. "Always beautifying the hotel or improving its lovely grounds. And what's this about you showing those newfangled moving pictures on the lawn? Won't you be the first hereabouts to exhibit them?"

Mr. Bobinger laughed heartily and the gold chain danced up and down on his big stomach. "Yes, indeed!" he boomed. "Always something new and exciting for our patrons. You must come soon to see 'The Great Train Robbery.' Since that is 5 years old, we're also featuring a brand-new and dramatic motion picture show: 'A Sailor's Return.'

See them both. After you have dined with us, of course, with your wife and daughter, Captain Clark." And Mr. Bobinger beamed, his chubby cheeks puffed up with pride and good humor.

"Agreed," saluted the captain. "And we'll bring our nephew, Michael Walsh, here. He's visiting us from England. Goodby!" Again he saluted, then signaled old John to hitch up the mule and proceed up the towpath.

The fairyland surrounding the handwaving hotelkeeper moved slowly out of sight. But not before Abby and Mike saw, in the settling dark, a fawn drinking from the creek. A mother deer stood near, watching.

Abby and Mike were so enchanted with the view that Mrs. Clark had to call them twice to dinner in the cabin.

A lighted lamp stood in the center of the table, on the oilcloth cover. The lamps bracketed to the walls were also lit. They were not at all as exciting as colored electric lamps, but their yellow flames were warm and cast cozy shadows about the cabin. Soon the Clarks, Mike, and old John, with his black face glistening in the lamplight, were all comfortably seated around the table laden with food, and their heads were bowed in prayer.

CHAPTER 4

How Cabin John Got Its Name.
Indian Relics and a Gold Mine.

Since Captain Clark did not have to deliver the *Daisy Mae's* cargo in Cumberland at any certain time, he decided to let old John take the day off to visit with relatives in the quaint community of Cabin John.

"And, meanwhile," he suggested to Abby and Mike, "let's inspect the gold mines near Falls Road and Conduit Road. That's not too far from Cabin John. But don't forget to meet us at Great Falls in the morning," cautioned the captain.

Mike's face was screwed up. He thought he missed something said by his uncle. "Did you say 'the Cabin of John'? Our old John?"

"No, son," chuckled the captain, "I said *Cabin John.* The town was perhaps named for your Captain John Smith, the English adventurer. Or for the Captain John's Run (or Creek), that flows past it. Or even for John of the Cabin, a mysterious hermit who once lived with his dog in a cabin on one of the hilltops above the canal. He fished in the creek and hunted all over these woods, where Indians had done the same for centuries before him.

"Some say that the old man was a pirate captain who hid away in Cabin John with the buried treasure he took from ships at sea and stole from his crew."

"I say," exclaimed Mike, his eyes bugging. "Then Captain John could have been a *real* pirate. But you called the place *Cabin* John."

"Well, I suppose down through the years people slurred the title of Captain into Cabin. But about that buried treasure — John Manson told me that anyone who buys certain property in Cabin John and finds any treasure buried in the ground there, must share it with the land development company that sold the lots. It's written into the deeds.

"Personally," he said, "I think it's a scheme to attract attention to the area, so that salesmen can sell off the land more quickly. Anyway, I daresay the digging exercise is good for the new landowners and good for the turned-up soil as well.

"Come on, kids — let's get along to the gold mines, where the digging for treasure is real!"

"Oh, no, you don't," protested Mrs. Clark. "Abby must first help me with the clothes-washing, or there'll be no fresh changes for any of us to wear."

She promptly put the girl to work over a washboard leaning against the inside of a washtub. Water, heated on the small coal stove, was dumped into the galvanized tub. Brown laundry soap was swished around, and soon the suds were heavy enough for Abby to soak the soiled clothing in them. She scrubbed and kneaded each garment over the bent tin ripples of the washboard. It was fun for awhile. So was the chore of getting out the sudsy water by hand-wringing. Then squishing and rinsing the trousers, shirts, dresses, stockings, and undergarments in clear water.

After an hour or so, however, Abby's arms and back grew tired. She was excused to join her father and cousin.

They already had both Bozo and Lazy Bones on a trail leading off the berm of the canal toward Seven Locks Road, Persimmon Tree Road, and beyond to Brickyard Road. Throwing a blanket over each mule's back, Captain Clark lifted his nephew atop Bozo. Tomboy Abby already straddled Lazy Bones.

Holding a rope attached to the bridle of each animal, the captain strode off, while the youngsters held onto the reins and pressed their legs against the barrel-like sides of the mules. Sometimes they balked, or stopped to munch on grass growing along the way. But their master tugged on the bridles and pulled, until the beasts decided to obey. Sometimes the party halted to pick and eat blackberries or elderberries from closeby bushes, or drink from a brook gurgling over smooth rocks.

The sharp-eyed captain spotted a muskrat, scuttling to hide under a wet rock. And two wild turkeys in the woods. He drew attention, also, to the pawpaw and several persimmon trees that grew near the canal.

"They both ripen in autumn," he said. "When the pawpaw skins grow brown, the insides taste almost like bananas. As for the persimmons," he winked, turning to Mike. "Your cousin learned long ago to pick them only when they became soft and mushy — after the first frost, usually; then they taste like apricots. Abby once bit into the unripe fruit and you could see her lips and mouth pucker up. She wasn't able to whistle or eat for quite awhile. Yessiree! Gave mother a good rest from preparing her dinner."

From a hilltop, they could look down on the canal lying snug against the Potomac River and turning and twisting with it off into the distance, through picturesque countrysides. Toylike cows and sheep grazed peacefully on the rolling pastures.

Turning deep into the woods, they seemed far away from any mark of civilization. It was easy then for Mike to imagine himself back in time — among the redskins that once lived all around there.

The mules approached a creek running downhill to join the Potomac. Bozo's hooves kept slipping on the sides of the bank. Abby, following on Lazy Bones, suddenly yelled: "Hey! Whoa there! Whoa there!" and slid off her mule.

Startled, her father turned to see Abby squatting near the creek's edge, up to her wrists in mud and dirt, clawing with her fingers at a bank of earth and roots. Mike dismounted and joined his uncle who had stooped and was watching Abby wash some objects in the creek waters.

"Father . . . !" Abby was breathless with excitement. "Look at these flint arrowheads and spear points. I watched Bozo's hooves kick them loose from that bank."

Captain Clark hunched over, picked up the wet objects and examined them closely. "Must have been an Indian village hereabouts," he mused, thoughtfully. "Probably the Susquehannas . . ."

"I say, Abby. That's quite a find," said Mike, fingering one of the stones. "I've never before seen or handled weapons used by American redskins. These are exciting. May I please have one as a souvenir to show off to my friends in England?"

"Perhaps Abby will let you take them all," suggested her father. Do you mind, Abby? You already have a good collection of Indian artifacts."

"Shucks, no — I don't mind," said Abby, although her eyes lingered on a spear point she would have liked to keep. She tucked the relics carefully into a knapsack on Bozo's back and the party continued on its way.

Soon, the tall, wooden shafts of the Maryland Mine Company came into view. The foreman, an acquaintance of Captain Clark, said they could visit the second level, 100 feet under the surface of the earth, where miners were chipping at gold ore in the sides of a great rock tunnel. The foreman handed out small kerosene oil lanterns and climbed down a ladder into a pit. The others followed, more slowly, for it grew dark and darker below them, with each step down the ladder. They lighted their lamps at the bottom and single-filed after the foreman, along a cool, black, dirt path that slanted downward into the ground.

"It's scarey here, isn't it?" whispered Mike to Abby. "It sure is," she agreed, gripping her father's hand. The group continued moving down into the darkness, following the light shed by their lanterns. Jagged rocks overhead and on the sides sparkled with mica.

More minutes of shuffling slowly down the curving caverns brought them to a number of brighter lights. Large, acetylene gas lamps, covered by glass shields, were set up where miners drilled into the rock. They hammered on chisels, turned them, then hammered again and again, until several round holes were formed. Then they carefully inserted sticks of dynamite into each hole. The foreman ordered everyone to retreat behind a protective rock wall of the corridor.

"And cover your ears. Tight!" he warned, sharply.

He lighted the fuse and raced for cover. A bright spark shone; it crept, spluttering, along the ground and up the wall . . .

A burst of flame! . . . A cloud of smoke. The earth shook underfoot. Crashing noises pounded around in Mike's handcupped ears. He felt ill for a moment, until he glanced up at his uncle and cousin and was calmed by their faces. At least they appeared calm in the half darkness.

When the dust and smoke cleared, the miners advanced and attacked the rock wall and its fresh, jagged, cavities with their picks, hammers, and shovels.

“Yippee!” one yelled. He fished both hands into a big hole and drew out a chunk of rock. “Just look at this, will you!” He thrust the quartz under the nose of the foreman, pointing to the surface of the rock. A yellow vein glittered, ending in a blob about the size and shape of a lima bean.

“A nugget, sure enough,” nodded the foreman. He was a cheerful, enthusiastic, man. He patted Abby, then Mike, on the shoulder, then stooped to poke through the fresh heap of rock chips. “You kids have brought us luck,” he said, “so here’s a good sample of gold-bearing rock ore for each of you.”

He escorted the party back to the surface of the earth. There he showed them giant rock-crushing mills. One had great jaws that ground large quartz rocks into small fragments. This helped release the tiny particles of gold they contained. (The process was like cracking walnuts, then picking out the bits of meat inside.)

Another way of breaking the large chunks of rock into small pieces was shown to the visitors: a huge round steel bin was loaded with gold-bearing ore and many steel balls, each as large as a tennis ball. As the bin revolved, the balls cracked and crushed the rock into bits.

Screens, belt feeders, and water sprays were used to sift, move, and wash specks of gold from the quartz pulp in the mills. The Clarks and Mike also watched flakes of gold appear on silvered copperplates that had been brushed with mercury. The poisonous quicksilver “grabbed” the yellow metal from rock mixtures.

The youngsters held their gold ore souvenirs tight in their fists and saw the “retort sponge” left in the furnaces, after the mercury was boiled off for later reuse. These small masses of gold were weighed carefully on a small beam scale, then placed in pouches made of chamois skin, for delivery to the United States Mint at Philadelphia. There they would be recorded and melted down into heavy gold bars. Eventually, they might be stamped into gold coins or made into jewelry.

As the party left, the foreman confided that some gold-bearing rock had been used in the original Conduit Road — the same “highway” that ran parallel to the canal and carried water in underground pipes from Great Falls to Washington City.

It was mid-afternoon when the Clarks and Mike returned to the *Daisy Mae.* The youngsters slid off the backs of Bozo and Lazy Bones and Captain Clark drove the animals across the mule fall-board. Mike led them into the stable and fed them hay and oats. He patted their heads gently and talked softly to them as they ate.

"You're a good hoggee, alright." Abby had poked her head into the stable and had been watching her cousin quietly for some moments. "Now come along; mother has made us a light snack."

Following their meal, the youngsters stretched out in the sun, for a nap, on blankets spread over the hatches. It seemed only a moment later that they heard:

"Wake up, you two! Let's go fishing before it gets dark."

Captain Clark had tied up the boat for the night at Widewater. The oldtimers called the place Log Wall, because timbers were used to form the base of the towpath. The captain also had cut three sycamore branches and, removing the twigs, had fashioned them into fishing poles. He secured a string and a hook to the point of each branch. Mrs. Clark furnished the bait — a clump of moist bread dough. The captain, his daughter, and nephew sat on the deck of the boat, their legs dangling over the bow, and their fishhooks deep in the canal water.

There was a tug on Abby's line. She yanked backward on her pole and the hook flashed up into the air — minus its bait.

"Yea-ah!" teased Mike. "Girls don't know how to fish!"

Just then his line went taut. Mike straightened up and put out his chest: he would show Abby how to catch fish. He played his pole back and forth until the end bent with the weight of a fish. Then he yanked the pole backward over his head. His line also had a bare hook dangling.

"Yea-ah!" mocked Abby. "Even English fisher boys don't know how to fish!"

By this time, Captain Clark had lured and caught three large catfish. They flopped about the deck, until Mrs. Clark gathered them up. She soon had them dipped in cornmeal and sizzling in lard in a frying pan.

Abby landed a sunperch; Mike caught a small carp. And guess who was kept busy cleaning the fish and cooking them for supper?

Abby ate her sunperch proudly, until a fishbone caught in her throat. It made her cough and Mrs. Clark pounded on her back until the bone was freed.

"You know you must always remove the bones before chewing on fish," she said. "Besides, Abby, you shouldn't have tried to out-brag Michael about your fishing skill. That caught bone may have been your punishment."

Abby and Mike spent the rest of that day reading copies of "St. Nicholas Magazine," until it grew dark and the night noises grew louder and made them feel sleepy.

CHAPTER 5

The Tavern and Great Falls.
A Savage Storm.
Fun Along the Canal.

Abby and Mike were awakened by the call "Hey-ey-ey, Lock!"

They heard the groans and squeals of closing and opening gates and the sound of rushing water. The *Daisy Mae* was being locked through. Mike rolled out of his hammock. He "hit the deck" and joined Abby in the cabin. They peeped out a window at the red Seneca sandstone walls creeping downward, stone by stone, until a large, impressive, white building appeared before them. It was the Crommelin House or Great Falls Tavern at Lock 20, a favorite inn for travelers since 1831.

Mrs. Clark saw the youngsters standing, barefooted, peering out at the two-storied inn with its great chimney, wide veranda, and shingled gables.

"Look at all those people moving about in their fancy dress clothes," said Abby.

"They look like city folk," observed Mrs. Clark. "Must be sightseers from far away," she said. "Perhaps New York, Philadelphia, or Baltimore. Many thousands come each summer from all around, you know, to visit the Nation's Capital. Some even come from 'way out West — or even across the ocean — like our Michael. And after seeing all the Washington sights, like the Capitol, the Monument, the Library of Congress, the Smithsonian Museum, the Zoo, and the Botanical Gardens — why, they come out here to oh!-and-ah! over the Great Falls of the Potomac."

A faraway look came into her eyes and she rested one finger against a cheek and said: "I remember my grandmother describing her honeymoon stay at the Crommelin House. She told me how the young ladies felt so grand in their wide hoopskirts, twirling their gaily colored silk parasols over their shoulders. They were escorted by beaverhatted young gentlemen swinging their canes. The couples would stroll arm-in-arm about the tavern veranda, my grandmother said, and along the canal, watching the mute white swans follow the boats, after they passed through the lock.

"That was before the Corps of Engineers built the intakes — you can see them on the towpath side of the canal, over there — so that water from the Potomac River could enter the underground pipes leading to the reservoirs of Washington City."

Mrs. Clark paused and the children watched her expression change as her thoughts returned to the present. She gave a short laugh.

"Here I am chattering away like our Captain Clark, when I only came in to prepare your breakfast. Hurry up now and get dressed. Old John's returned from his overnight in Cabin John and is helping to dock our boat out of the way of passing barges, while we show off the Falls to Michael."

Already, the motion of the *Daisy Mae* slowed to a halt, and they could hear the rustling, soft scrape of ropes being secured and the cheerful banter between Captain Clark and John Manson as they worked.

The youngsters washed up in the basin, dressed, and hurried through breakfast. They soon stood in front of the Inn's candy showcase where each exchanged a cent for peppermint sticks to eat later.

Captain and Mrs. Clark were waiting on the towpath, in front of a swinging bridge with a big sign at the entrance reading "5c Fare." Supported by steel cables, the wood bridge was really only a plank floor with wire sideguards. Long and narrow, it snaked away over the rocks, toward the falls, wiggling from side to side and swaying up and down, as people walked over it. The bridge looked scarey. But exciting.

Captain Clark paid the fares and they all moved gingerly over the bridge, high above foaming white, bubbly, water that roared over and around submerged rocks. Abby and Mike kept a tight grip on the sideguard cables until they had crossed the bridge. Stepping onto a dry area where great boulders were scattered all around, Captain Clark called attention to some rock tops.

"See those round potholes in them, Mike? You could hide inside that big one, over there. Those holes were ground out by pebbles, whirling 'round and 'round for hundreds, maybe thousands of years." Mike went over and felt the inside of the rock with his hands. It was smooth and round. Then he climbed over the side and sat down in the hole, with only his head and shoulders sticking out.

Just then, an alert tourist stepped up with his Brownie box camera and clicked its shutter almost in Mike's startled face.

The boy scuttled out of the rock hole and ran to the Clarks. "Ho! Ho! Haw!," laughed the captain, while Mrs. Clark and Abby tried not to. "Sorry, Mike — but, by golly, you did look like a hen frightened off its clutch of eggs." He began laughing again and slapping his thigh as he doubled up. "Won't that fellow have fun showing off your snapshot back home in Weehawken, Tuscagoula, or some such place."

Mike felt embarrassed and was shy for awhile. But Abby grabbed his hand and yanked him across a small wooden bridge or platform that led over a cleft in the rocks, where the Potomac raced swiftly "away" from beneath them. "Feels like we're on a boat and moving in the opposite direction, doesn't it?" observed Abby, smiling at her cousin.

It was an optical illusion of movement, of course. But it *was* unnerving to see all that water surging so close underfoot, hear its deafening roar, and feel the spray filling the air and wetting their faces and clothing.

Captain Clark strode to a hill; he bent and poked around in it. "Mike! Abby!" he called. "Come over here, please." In the palm of his hand lay a number of tiny white sea shells. Each was smaller than the smallest fingernail.

"These shells prove," said the captain, "that the rocks on which we're standing were once underwater. Would you believe that a deep sea bottom was here, ages ago?" Mike stood silent, looking all around him and marveling. He imagined himself walking on the sea bottom among the fish, with a giant sea turtle swimming overhead. Then he remembered he had to breathe. And he was not even wearing a diving suit. Not even a bathing suit. He snapped his mind back to "safety" — like waking oneself from a bad dream.

He hurried to join the Clarks. They were walking down a path between huge boulders that made them look like dwarfs. They all turned past the last rock and there they were — viewing the river valley.

Captain Clark waved his arm to the right, like an actor drawing aside a curtain. "Intro . . . ducing the Great Falls of the Potomac . . . ta-ra-da-da . . . !"

As far as they could see upriver, monstrous rocks were strewn all around, as though by supernatural hands playing some gigantic game. And water surged in white fury at and over the rocks, flinging lacy curtains of spray high into the air. Water rushed and whirlpooled angrily. Green-blue water turned to white as it smashed fiercely at the rocks in its path — then zoomed skyward in its rage — thundering its frustration.

"What a roaring noise," protested Mike, clapping his hands over his ears.

"And there's a rainbow across the river." Mrs. Clark was pointing to where a rift in the clouds projected the sun's rays. "Over that cloud of mist churned up by the falling waters."

"It's awesome, that is," gulped Mike.

"Gosh-awful scarey, too," said Abby. She turned and climbed easily up the side of a boulder that was as big as a house. It was covered with green lichen. When she reached the top, Abby sat down and admired the falls. "Come on up, Mike," she beckoned. "It's fun up here."

Mike was awkward as he climbed slowly up to join her, choosing a route over smaller, nearby, rocks. As he drew closer to the top, he leaped suddenly from one rock to the top of another. But this one was smooth and Mike's shoes slipped off the side. Down he went, crashing out of sight into bushes growing bravely in a small patch of earth between the rocks.

"Help!" he shrieked from atop a heap of twigs — evidently a squirrel nest.

Abby looked around, frightened. She saw nothing but rocks.

"Where are you? Mike, where are you? Keep calling so I can find you."

Twigs were sticking into Mike's back and the rocks were too close together to turn around in and stand up. He lay there, croaking: "Help! . . . Help!" Suddenly, there was Abby's face looking down into his over the edge of the rock. Mike looked awfully white and scared to her. And silent. She reached down and Mike grasped her hand and lifted himself slowly, carefully, off the mass of bushes and twigs.

"Umph! . . . U-m-m-ph! . . ." Abby grunted, straining to help lift Mike up the side of the rock. They were both wet with perspiration and shaky, when Mike finally scrambled slowly, but safely, over the top of that rock and down — very carefully, with each step firm in a rock crevice — until they reached the earthen path. Abby found her parents still standing, fascinated by the falls. Abby would not tell them where they had been and what had happened to them, but Michael blurted out everything — so frightened was he.

His folks quieted him down, but Captain Clark was upset, "Well, young man," he said, sternly, fingering his nephew's torn clothing. "You must be trying to establish some sort of record for recklessness."

"Look at his poor scraped hands and elbows," lamented Mrs. Clark.

"But aren't you glad he didn't break an arm or leg?" said Abby.

"It's a wonder!" agreed her mother.

They half-carried, half-led, the bruised boy down the path, back over the swaying bridge, and to the dock near the Great Falls Tavern. There, on a box, one leg crossed over the other, sat old John, calmly smoking his pipe. His jaw dropped — and so did the pipe — when he saw the sorry-looking group in front of him.

Captain Clark tied up for the night in a quiet spot along the berm, before reaching Swains Lock — lock number 21.

During most of that night, Mike tossed in his bunkhouse hammock. He was sore and overexcited by his bad experience at Great Falls. And in his fitful dreams he seemed to hear again the frightening, rushing, waters of the falls and felt again the swaying and lurching of the wire-suspended bridge.

He awoke suddenly. He heard loud shouting. His uncle was on deck, yelling orders to old John. Mike heard the wind screech louder and louder and through a window he saw the dark, trembling treetops. Whipped and lashed by the heavy rain, the boughs swayed lower and lower over the canal.

All at once, Mike saw a great branch snap off and fall. C-R-A-S-H! S-P-L-A-S-H! It hit the water beside the *Daisy Mae.*

Mrs. Clark awoke, screaming. And Abby jerked herself upright in her bunk. A wave slushed over the water-rail. It ran the length of the raceway around the hatches, washing down the steps into the cabin, where Captain Clark had returned to struggle into his boots. He said something sharply under his breath.

"Sorry, mother," he called, quickly, to his wife. He made his way to the bunkhouse, plucked Mike out of his hammock and carried him down into the cabin.

"Keep the kids in here," he ordered. "Safest place for them." And he was gone, hurrying ashore to help old John secure the bowlines through ring bolts, to a tree by the towpath. Though the boat rocked ominously back and forth and up and down — just like the suspension bridge at Great Falls — it now seemed more safe.

Abby and Mike stared through a rain-streaked window. Whenever flashes of lightning zig-zagged across the sky, they could see dark shapes move about swiftly on shore. Mrs. Clark came over to comfort them. "We'll be alright," she said, putting her arms around them.

Captain Clark was now in the stable at the bow, patting the heads of the mules and gently slapping their necks and shoulders, for they were braying and stamping their hooves in uneasy fear.

All at once: a crash of thunder. Then a fierce ZAP! sounded outside, followed by a brilliant spear of lightning. The heavens lighted up and the rain-splattered window turned glaring white. The eyes of the children were blinded. The *Daisy Mae* jolted to one side, swayed, then began moving slowly, then faster and faster, as though trying to escape from the thunderclaps and the lightning.

Mrs. Clark shrieked her fright and fearing for her husband, ran up on deck. The captain was heard shouting: "Flop on the deck and hold fast to something." Then he roared down into the cabin, over the whine of the wind: "Stay in the cabin, kids. That's an order!"

He was yelling to old John, who was apparently on the dark towpath. "We're loose! The old tree's been struck by lightning. Toppled into the canal; ripped loose our lines."

The youngsters felt the barge rock and turn first one way, then another.

A scream rose in Abby's throat, but she managed to stifle it: she feared that her mother, lying flat somewhere out there on the dark, wet, deck, would become frantic. Mike's trembling hand reached out and grabbed Abby's and held it tightly . . .

The stern surged up, then down, up again, almost tossing Abby and Mike off her bunk where they sat, onto the cabin deck. Once they felt lifted completely off the straw mattress and grunted "Oof!" when the barge came down with a loud THUMP! on the water. Then they heard loud shouts above the hard-pelting rain and the roar of the wind. The barge rocked and creaked and groaned as though in agony; it swayed from side to side, struggled to squirm from the grip of the storm.

The force of the wind drove the stern hard against the bank. The hull scraped and screeched.

Inside, on the cabin deck, lay a tangled mess of bedding and children.

Knocked to the floor, Abby and Mike remained quiet for a moment, afraid to move. Then, forgetting bruised backsides and danger, they held onto each other in the dark and crawled on hands and knees up on the deck. Out into the driving rain. There, in the dim light, they saw the Clarks and old John. They had scrambled aboard the boat and were clutching the sides of the hayhouse on the sloping deck of the boat.

Mrs. Clark was the first to see them. "Children!" she called. "Children! Crawl over here."

The wind and rain tore fiercely at the soaked nightclothes of Abby and Mike. But they kept crawling along the raceway and gasping and gulping air into their lungs, like half-dead fish. Holding onto the corners of the hayhouse, the Clarks and old John reached out and dragged the children across the deck toward their own sheltering bodies. Abby and Mike lay still, face down, fighting off the sobs rising in their chests. But now they were warmer and shielded from the rain by the forms huddled over them.

Whenever lightning flashed, the stern of the *Daisy Mae* could be seen resting on the bank. Also the canal — littered with broken tree branches. These poked up from beneath the angry masses of water that rolled over and swirled around the limbs.

"The mules are sure kicking up a fuss," said old John hoarsely, lifting his head and listening to the frantic hoofbeats coming from the stable. "Can't stand it. They must be awful scared. Gotta help 'em." He shifted his body off Mike and made his way to the stable.

The mules were rearing, their eyes big with fright. Old John spoke softly to them and patted their heads, necks, and flanks until they quieted down.

The Clarks and Mike felt the pelting, stinging rain slow gradually into a drizzle. Then it stopped. The choppy waters in the canal grew calmer. The strong, loud wail of the wind thinned to a whisper . . . Suddenly . . .

All was quiet.

Overhead, the low, dark clouds boiled apart, revealing the bright twinkle of distant stars. And then . . . the round, full face of the moon — a glowing lantern in the sky.

The storm had passed.

Four figures half-rose off the soggy deck. They clung to the bunkhouse for support, for the deck was slanted. They squinted around in the moonlight. Old John came crawling from the stable, like a tired crayfish. The captain muttered "Be right back; stay here" and moved slowly, carefully, across the tilted deck to the cabin. He found two lanterns and lit them with matches he took from a watertight container deep in his drenched trousers. Aided by the moonlight and the lanterns, he made his way back to Mrs. Clark and the children.

"Here, darling," he said to his wife, handing her a lantern. He hugged her and, smiling, patted Abby and Mike. "Must check the damage to the boat," he said. "Meanwhile, get over to the cabin and make yourselves as comfortable as you can." And holding the other lantern, he and old John crawled away across the inclined deck and over the aft rail of the boat.

They found the rudder and stern badly scraped, but not cracked or broken. "The old *Daisy Mae* came through just fine," said Captain Clark, quite pleased. "We'll wait 'til daylight; then we'll have her dragged off the bank and back into the canal where she belongs. Meanwhile, let's change into dry clothes and bed down for a bit of sleep. I'm plumb tuckered out and you must be too, John."

Captain Clark awoke at dawn from his short, bad-dream, naps. Lying flat, he twisted his aching body, crablike, across the sloped deck. He rolled over the water-rail and stood erect on the towpath. "Ah . . .!" he breathed deeply, throwing out his chest, "A pleasure to stretch." He looked all about him. Debris was scattered everywhere.

Examining the *Daisy Mae* more closely in the light, the captain decided it had suffered little from the storm. We're lucky, he thought. Only scraped the underside on the earth and brush of the bank. We'll have the hull sanded down and freshly protected with paint and pitch when we get to the boatyards at Cumberland. He was grateful for his good fortune and the safety of his family. And I mustn't forget, he concluded, happily, that John and Lazy Bones and Bozo are all unhurt.

The captain called to other, less cheerful, boatmen nearby. They were shuffling around, clucking their tongues, and shaking their heads at the destruction all about them.

Captain Clark gathered some rocks from the woods and fashioned a crude fireplace on the edge of the towpath. He returned to the *Daisy Mae* for some dry wood and coal. "Come, mother," he said to Mrs. Clark, "Let's get ashore and heat water for soup and oatmeal to serve the children. And to the others out there less fortunate than we."

Abby and Mike felt safer once they slid over the boat's side and felt firm ground underfoot, even though it was still muddy.

"May we go exploring?" asked Abby, gulping down her last spoonful of oatmeal.

"You may," said Mrs. Clark, cautiously, looking about her, "if you promise not to bother anybody. All those poor souls around here are trying to set things right and get back to moving up and down the canal. I must go over to them with pots of hot oatmeal and coffee. But aren't we lucky! Only a few pieces of chinaware broken last night."

"Yes, mother," agreed Abby, "plenty lucky. We'll help you straighten up things," she called over her shoulder, "when we get back." She grabbed Mike's arm and drew him down the towpath.

They stared curiously at the smashed cabins and hatch covers of some barges. Trees, uprooted from the bank, had come crashing down on the boats. Some had stove-in hulls; the raging canal waters had driven them, helpless, against rock walls that bordered the canal at certain points. All sorts of debris floated, half-submerged, on the water: chairs, bedding, wooden boxes and crates, stools, straw mattresses, buckets, tubs, feather quilts, papers, bottles, jugs, and pots.

Soon, Abby and Mike grew uncomfortable watching the dazed canallers and their children wandering around, trying to sort out their belongings. They mumbled what they hoped were cheering comments, and returned to the *Daisy Mae.*

They watched Captain Clark and old John at work. Several teams of mules had been borrowed and harnessed together. Some canallers also helped secure stout towing lines around the hull of the stranded boat.

"Stand clear!" warned the captain. "Stand clear! We're ready to heave 'er back where she belongs."

"Giddy-ap! Giddy-ap!" yelled old John, snapping his whip over the heads of the mules. They strained their necks into their collars. The *Daisy Mae* groaned and squealed as she was turned slowly over onto her keel. She crept, creaking, down the bank of earth and underbrush.

SPLASH! She slid into the canal and sent a wave of water surging upward from beneath her stern. Down fell the water on Captain Clark and old John, drenching them. But they laughed aloud; their boat was safely afloat.

Abby and Mike joined happily in the laughter. They jumped aboard to help tidy up the hold, where they found the cargo shifted and tumbled about, but not damaged.

When Abby had finished in the hold, she helped her mother set the cabin right. Then she came up on deck and into the sunshine. The sky was blue. The air was clear and warm. Mike was sweeping out the hayhouse. He grinned over at his cousin and gestured at the towpath. Beside it, strange canallers sat in the weeds, downing the warm breakfasts their wives had prepared over Captain Clark's crude fireplace.

Soon, up and down the canal, everyone was busy getting things in order. Some canallers began singing "Carry Me Back to Old Virginny," as they worked, sawed, and hammered. They heaved and tugged, strained and pushed, pulled and hauled, until the canal waters were cleared of submerged tree limbs and some of the battered boats could be repaired.

All that day and the next, the men helped the Canal Company work crews clear debris from the waterway. Later, everyone rested. The women put fresh, clean bandages on the scratches and cuts, and torn skins of those who had been injured; they cooked and baked while their children helped set up some makeshift tables on the towpath for a grand picnic.

One of the older boys was a prankster; he was surrounded by several younger children. He showed the group a red-and-white peppermint candy stick he held in one hand. Smiling, he said to the little boy standing before him:

"Open your mouth
And shut your eyes,
And I'll give you something
To make you wise."

The tot obeyed and the big boy shoved a redhot pepper into the open mouth of his victim. The poor little fellow raced off, screaming, to his mother, with tears streaming from his eyes. The group (they had already been "suckers") laughed and moved off with the trickster to another part of the towpath, looking for fresh "dummies" who were not watched carefully by parents.

"Cruel trick, that," said Mike.

Abby grinned. "I thought so, too, the first time I got fooled into biting on that hot stuff — when I was a little girl."

They watched some young canallers start a square dance on the towpath. One produced a "squeezebox" or accordion; another started scraping away on his fiddle. Old folk began clapping their hands in a steady beat. A lean captain's wife used an orange crate to hold her baby, while her husband lashed their 2-year-old son to a tree with a piece of harness leather. The little fellow nibbled calmly on a carrot and watched his parents take their place and begin whirling through the "Virginia Reel." Though they were tired and not dressed for the occasion, others joined in, inspired by the magic of the music. They sashayed 'round and 'round, in gay forgetfulness of their recent troubles. These were past for awhile and life was again good along the canal.

Soon, even the young people got into the spirit of things. Abby and Mike got into a single-file line that snaked up and down the towpath, yelling more than singing:

"Dan, Dan, the funny old man —
Washed his face in a frying pan,
Combed his hair with a wagon wheel
And died with a toothache in his heel."

Abby led the line in dancing and singing more nonsense verses, until Mike felt that he had laughed himself silly and was weak and exhausted. They begged off and went to join Captain and Mrs. Clark, who were sprawled on blankets spread on grass in a quieter spot off the canal towpath, together with older captains, their wives, and crewmen. The men puffed on their pipes and swapped stories; they bragged and told tall tales. The women gossiped among themselves. Occasionally, all would stop for a few moments to heehaw together over memories about some of the folks who once lived and worked on the canal or near it.

"There was this captain's wife," recalled one fat boatman, removing his pipe from a hole between two front teeth in his lower jaw, "who had a really fast dog. 'Twas a rabbit-hunter and it would chase after and catch the little critters — on the run, too.

"Well, sir, whenever this here lady spotted a rabbit on or near the towpath, she'd toss her dog overboard in the direction of the cottontail. And water or no, that hound would zo-o-o-m!, just like that (he mimicked the dog's speed by slapping his hands together, then sliding them swiftly, one past the other), over the water, up the bank, along the towpath, and into the woods. Bingo! He'd catch Mr. Rabbit . . .

"My-oh-my! That captain's lady always had a kettle of rabbit stew aboilin' in the galley. And if'n her dog caught too many at one time, why she'd trade off them rabbits or sell 'em to the bargemen."

"Yes, indeed," said Captain Clark. "Most canal folk have been mighty generous to one another." He reminded his listeners of "Pinty Wine" and his wife, Maggie. Long ago, they were in charge of a work scow at "Falling Waters" near Williamsport.

"Remember that fine couple? They were never satisfied unless they were giving something to other people. Things like milk, butter, fish, vegetables. Sometimes, even a warm, fresh pie. A good, canal-working friend of mine, George 'Hooper' Wolfe by name — that's the character (he looks like a chubby Irish leprechaun) sitting with his back to a tree — told me yesterday that when Pinty and Maggie died, God must have called them 'home' because He needed a good pair of helpers."

The oldtimers nodded their appreciation, remembering "Pinty Wine" and his Maggie.

"Hooper" Wolfe got up, slapped the dirt off the seat of his pants, and strolled over to put his hand on Captain Clark's shoulder.

"Cap'n," said he, "I've been called all kinds of names in my time, but never 'Irish leprechaun.' Be that a fightin' word, professor?," he teased. "With my fourth-grade education — why don't we just settle on 'elf,' hey?" And he joined in the laughter at Captain Clark's expense.

"But," said "Hooper," holding up his hands for silence, "many canallers have answered to animal names like mine, for instance, of 'Hooper' Wolfe." And he ticked off on his fingers "There's Captain Daise Wolfe, Harry Coon, West Bear, Tom Mouse, and Charles Fox." He scratched the whiskers on the chin of his jolly face. "But, you know," he chuckled, "I never met or heered tell of any Captain Ratt."

To their delight, Abby and Mike learned that many canallers around them had colorful and unusual nicknames. There was Uncle Punko and his wife, Auntie Whozat. He called her that, because she always asked "Who's 'at?" when she spied some stranger. The youngsters also were introduced to Pickle Face, Frogface, Thingamajig and Thingamabob (twin brothers), who were related to cousins named Rinkeydink, Toothless, Sputter, and Zippo.

At that point, Uncle Zippo turned to Mike and embarrassed him by telling the folks how the lad, following a hot day's work spent clearing debris from the canal, had gone swimming.

"Mike here was thrashin' the water mighty close to the berm," said Uncle Zippo, "when I spotted a snapping turtle near him, hidden by a floating tree limb laden with leaves. Mike, I figgered, might lose a finger, or toe, or mebbe get a chunk of backside chomped outa him by that turtle, so I bent down right fast, lifted the limb outen the water and brought the leaf-free end down right smart on the back of the varmint, cracking its shell open.

"That there turtle was soon simmering in my missus' soup kettle. Sure was delicious," said Uncle Zippo. He rolled his tongue around his lips in happy memory. "But you know one thing? — li'l ol' Mike here — he wouldn't taste none of it when I axt'd him."

Everyone laughed, but poor Mike's cheeks reddened. He moved to the back of the crowd. There he was joined by Abby and the two walked down the towpath; every once in awhile they paused to skip

stones over the canal water. They also stopped to watch some boatmen and their families gather driftwood to store on their boats for later use in cooking meals. Boatmen headed for Cumberland scraped and swept their barge holds clean of leftover coal for their stoves. Their wives preserved fruits and vegetables, bought potatoes from nearby farms and stored them. Other boatmen fished, or trapped wild turkeys, coons, or deer.

"Hooper" Wolfe must have been trailing behind Abby and Mike, for he arrived holding a watermelon on his shoulder, chanting as he strode along: "Red watermelon — red to the ri-i-n-d!" He cut generous slices for everyone who crowded around — including Abby and Mike. Mike had never eaten watermelon. But he watched his cousin happily devour her slice. In moments, he too was spitting out the seeds to distances almost as far as Abby's and the red juice was slobbering down his jaw and over his hands.

Abby kept some of the wet, slippery, melon seeds in her hand. She showed Mike a certain way to grip them between the fingertips. Zip! the seeds flew up and through the air and — plink! — into the canal.

"Hooper" showed Mike how he made his favorite food producer: a rabbit snare. He used his jackknife to cut a strong twig. He bent it over into the ground with a rope loop and a wood trigger fixed in just the right place. The bait was a piece of apple.

The canaller, who had been a barefoot muledriver when he was Mike's age ("couldn't afford no shoes then") said that he also put out rabbit boxes to catch the animals during his trips up and down the canal. These boxes would be propped up on one end by a small stick. Came a victim, the stick would be knocked aside and down came the trap.

"Of course," said 'Hooper,' "sometimes a young chicken got itself caught this way . . ." He winked. ". . . But I never heered anyone complain."

Abby confided to Mike that some youngsters she knew were always "finding" a loose, squealing pig, or a squawking chicken, or fresh eggs on the farms that bordered the canal. If the "finders" were caught redhanded by the farmer, they brought disgrace on themselves and trouble to their parents. Otherwise, said Abby, there were a few and quick, but delicious, barbecues and chicken dinners and omelets — with many innocent guests called in suddenly to share the ill-gotten goodies.

"But father warned me he would 'tan my hide' if he ever caught me doing such things," said Abby. "And I didn't dare ask him whether he meant sneaking around to catch a pig, or chicken — or eating them, afterward."

DAISY MAE
HARPERS FERRY, W.Va.
FROM SANDY HOOK

CHAPTER 6

Ghosts at Haunted House Bend.
Point of Rocks, Harpers Ferry,
and Antietam Battlefield.
Rumsey and the First Steamboat.
Civil War Raiders.
Andy Jackson Hears Some Old "Music."

It came time to get the mules on the towpath, hitched to towlines. And the mended barges followed each other — either up the canal toward Cumberland, or headed down to Washington City.

The *Daisy Mae* rose higher and higher. Through Swains Lock, Pennyfields Lock, and Violets Lock. It rose still higher through Rileys Lock and crossed the marvelous Seneca Creek Aqueduct. Such aqueducts looked like bridges, but their tops were filled with canal water so boats could float over them. Canallers could look down and see broad stretches of the Potomac revealed in all its beauties. They could also see the fishermen and small boats, some distance to the left, on a wide expanse of slack water.

"Say, Mike," said Captain Clark, "do you remember the Union Arch at Cabin John? . . . You do? Good! Well, that's Seneca . . . over there," he pointed. "It's famous for the reddish-brown sandstone in its quarries. It was used to build this aqueduct we're crossing and the copings atop the Union Arch. Yessiree! General Montgomery C. Meigs, the U.S. Army's great engineer officer, ordered those blocks boated down the canal. He also used that Seneca stone to build the Pension Office Building in Washington. Inaugural Balls were held there for Presidents. Something like Coronation Balls in your country, Mike.

"And the Smithsonian Institution building you've been to in Washington, Abby — you know — the museum that looks like an old Scottish castle . . ."

"Or like an old English castle," interrupted Abby, glancing at her cousin.

" . . . Right," agreed her father. "Seneca sandstone was used to build it. Also parts of the Library of Congress and the Capitol Building (where General Meigs put up the cast iron Dome), and even the backup stone of the Washington Monument. Meigs sure liked to use that Seneca stone . . ."

As the *Daisy Mae* came within sight of Harrison Island, one of the largest in the Potomac River, Captain Clark looked mysteriously at Abby, then at Mike. He fumbled with the tiller, took off his cap and mopped his brow with a blue bandana handkerchief. At last, he said, hesitantly: "Did you know, Mike, that superstitious canallers never tie up their boats or their mules at 'Haunted House Bend' — over there on the left?"

"How could he know that?" broke in Mrs. Clark, "He's never before been 'round these parts . . ."

"Figure of speech, my dear," said the captain, lightly. "Figure of speech. You see, Mike, 'Haunted House Bend' was one of the best locations to tie up overnight — 'til the Civil War and the Battle of Balls Bluff. That was fought over in Virginia — on the other side of Harrison's Island — about a mile straight west of here, I'd say.

"Our famous Supreme Court Justice, Oliver Wendell Holmes, was wounded at Balls Bluff. 'Tis said that when Holmes became a colonel, he guided President Lincoln through the battleline near Washington. Wearing his tall, stovepipe hat, the President exposed himself to enemy fire, whereupon Holmes shouted: 'Get down, you damn fool!'

"Anyway, canallers now fear the 'Haunted House Bend,' because it's so close to Balls Bluff and its small national Civil War cemetery. Many Union and Confederate soldiers rest there in unmarked graves . . . until nightfall . . ." Captain Clark paused for a full minute. Suspense and curiosity grew almost painful, until . . .

"What happens at night?" blurted Mike.

"Well," sighed the captain, heavily, as though he really didn't want to talk about it, "the ghosts of the unknown soldiers rise up — so the boatmen say. They float through the air from Virginia, across Harrison Island, across the Potomac, and come to earth at 'Haunted House Bend.' There, they scare mules, untie boatlines, and cast barges adrift. Some captains swear such mystifying things actually happened to them and their crews. The mules wouldn't even eat their grain; they were that scared.

"But *I* don't know, kids . . . It sort of reminds me of what Shakespeare's Hamlet said — you and Abby will be reading it in your English literature classes, I'm sure: 'There are more things in Heaven and Earth, Horatio, than are dreamt of in your philosophy.' "

That night, Mike tossed in his hammock, uneasy over the boat's nearness to "Haunted House Bend" and all those ghosts. Especially when the noises of the night began sounding louder all around: the baying of dogs on the farms near the *Daisy Mae,* the shrill whispers rising and falling over the meadows and swamps, the mournful wail of a B & O freight train . . . Everywhere, invisible bullfrogs and hoot owls and katydids vibrated the air until dawn, in a thrumming, whirring trap for the boy's troubled dreams.

They passed Whites Ferry, where Confederate General Jubal A. Early forded the Potomac to march on Washington in 1864. Generals Robert E. Lee and Jeb Stuart also had crossed their armies nearby, during the Civil War. About 2 miles further on, they floated near a quarry. "From there," said Captain Clark, happy he had another historical tidbit to share with his polite, attentive, nephew, "came the limestone columns used in the Statuary Hall of the U.S. Capitol."

The *Daisy Mae* was towed across the Monocacy River Aqueduct. Built of white granite quarried on Sugarloaf Mountain, several miles to the east, the span of seven arches was the longest, most attractive structure on the entire C & O Canal, declared Captain Clark. The marvel was built, he said, mostly by Irishmen brought to America just for this task.

Point of Rocks came into view, and hearing a noise, as of thunder, Abby and Mike glanced up, startled, to the right. There came a locomotive of the Baltimore and Ohio Railroad, drawing its long tail of passenger and freight cars. The train rushed along the narrow strip of land above the canal. It belched smoke and steam. Chuff! chuff!, it came, chuff! chuff! chuff! — clanging its bell and blowing its whistle. It hissed steam, puffed cinders, black smoke and soot into clouds that hung over its smokestack. It screeched away, over the rails, as though trying to drown with noise the silent barge, below it on the canal. The frightened mules brayed, trying to match the horrifying sounds coming from the monstrous "Iron Horse."

Every 15 miles or so along the canal, the *Daisy Mae* passed section crews stationed on work scows laden with rock shale, cement, gravel, straw and boards of rough lumber.

A level walker (foot patrolman), assigned to the crew, would each day examine his section of the canal. He was always looking for trouble, because in its heyday the canal had to be kept open for 800 boats, pulled by 3,000 mules towing almost 100,000 tons of cargo.

Mostly, the level walker would check the banks for leaks caused by colonies of muskrats. These little animals burrowed their nests deep into the muddy canal walls. Such disturbed "dams" caved-in easily. Were the holes left to grow larger, water would burst through the bank and escape, leaving boats stuck on the canal bottom.

To prevent such accidents, work crews scurried around with large wooden scoops or wheelbarrows, filling these with patching materials and clay scrounged from a nearby field. Such emergencies happened often, and a farmer with a ripening cornfield next to the canal, could awaken one fine morning to find that his crop had been "borrowed" during the night to help close a gap in a bank riddled by muskrat nests. Of course, the farmer was paid for water damages and the cornstalks stuck into the canal bank to strengthen it. But not for his turning the air blue with his opinion of the weary, "thieving" work crews.

The barge, for persons viewing it from a height, became a part of the inspiring view at Harpers Ferry. There, the Potomac River cut through the Blue Ridge Mountains to join the Shenandoah River.

Captain Clark tapped Mike on the shoulder to get his attention.

"High on that hilltop," he said, pointing, "is a famous rock. Thomas Jefferson stood on it about a century ago for a better view of 'one of the most stupendous scenes in nature . . . worth a voyage across the Atlantic.'

"We'll get up on that same boulder, Mike, before the summer's over. Then, I'm sure, you'll agree with Jefferson that the sight will have been worth *your* trip across the Atlantic."

"Jolly good," said Mike. "I shall look forward to it." Then he frowned. "I beg your pardon, uncle, but wasn't Mr. Jefferson one of the Revolutionaries who angered King George III?"

"Indeed he was that patriot. He drafted our Declaration of Independence from Great Britain in 1776, on the Fourth of July. Twenty-five years later he became our third President; 25 years still later he died, exactly on the Fourth of July, 1826. Jefferson was a man of many talents: inventor, architect, violinist, farmer. Above all, he was a truly great statesman.

"At Harpers Ferry, Mike — soon after we won our freedom from your country — an arsenal was built. It produced many of the flintlock rifles used in the War of 1812." The captain started to add "when we licked you British again" — but decided not to rub it in.

Instead, he continued: "And high up there on the left, among those buildings — you can barely see it through those trees — is a fire enginehouse. It was used as a fort by the mad John Brown and his sons and followers, almost 50 years ago. They raided the Federal arsenal at Harpers Ferry for guns and ammunition to use in freeing slaves by force. Government troops, led by the future General Robert E. Lee, captured Brown and most of his men, who succeeded only in hastening the Civil War. Harpers Ferry was considered the 'key' to the safety of Washington City during that war and the arsenal and armories were destroyed to keep them from falling into the hands, strangely enough, of the very same General Lee who had once protected the town."

As the *Daisy Mae* crossed the Antietam Creek Aqueduct near Sharpsburg, Captain Clark reminded Abby and Mike that Miss Barton there experienced the "Bloodiest Day of the Civil War," with a toll of almost 25,000 men killed or wounded.

"Abby, you'll be reading about Antietam in your history books," said her father. "About 'Bloody Lane' and 'Pack Horse Ford,' about the Dunker Church and Burnside Bridge . . . under it, Antietam Creek ran red with human blood. And you'll read how Miss Barton, while nursing the wounded, felt a bullet fly through her clothing, killing a soldier she held propped in her arms."

Above Lock 37, on Antietam Creek, were the ruins, the captain said, of an old furnace that had produced iron for cannon balls used in the Revolutionary War. This furnace was probably built on land sold or traded to Colonial settlers by Indian Chiefs of the Five Nations. They measured off pieces of land by "shoots" — the distance an arrow could be shot from its bow: 200 shots from the mouth of the creek, then 200 shots up the Potomac — and so on — squaring off the property.

And passing through Lock 38 at Shepherdstown, Captain Clark told Mike that the home of James Rumsey, a blacksmith, tavernkeeper, and friend of George Washington, was once nearby.

"In 1787," he explained, "Rumsey built and ran his 'Flying Teakettle' — the first steam-propelled ship in the world. It wheezed and puffed up clouds of steam and smoke. But it inched upstream at 4 miles an hour, against the swift current of the Potomac River. That was 20 years before Robert Fulton — did you know he was really a painter? — sailed his paddlewheel steamboat, the *Clermont*, up the Hudson River of New York.

"And, Mike . . ." he turned to face his nephew, "You'll be interested in this: Rumsey later went to England with a perfected model of his steamboat. While lecturing there to a distinguished group of scientists and engineers, he suddenly fell dead. He was the very first American to be buried, with full Parliamentary approval, in Westminster Abbey."

"Thank you, uncle," said Mike. "That was quite interesting information. But since Rumsey's 'Flying Teakettle' and Fulton's *Clermont* were invented long ago — why aren't steamboats used on the C & O Canal today — instead of poor towing mules like Lazy Bones and Bozo?"

"A canny question, lad," said Captain Clark, nodding his head. "Small steamboats do operate on this canal as passenger packets, excursion boats, or sometimes as grain and coal carriers to Georgetown. We just haven't met up with one on this trip. Yet." He frowned and his bushy eyebrows came together in a hard line on his forehead. "Steamboat propellers churn up the mud on the canal bottom and weaken the earthen banks. Besides, those boats make an infernal racket and often move much faster than the 3-mile-an-hour speed limit. I wish the Canal Company would bar them from this quiet and beautiful waterway."

The captain seemed sad and remote for some moments. He studied the peaceful surroundings. Then he sighed and commented to Mike that above Lock 39, nearby, was the Killiansburg Cave. "That's where women and children hid for safety, during the Battle of Antietam," he said. "Think of it . . . women and children . . . huddled together in fear . . ."

The *Daisy Mae* was 92.5 miles from Washington City and midpoint on the C & O Canal, when it neared Lock 43, below the village of Falling Waters.

"There," said Captain Clark, pointing out the spot to Mike, "was where our War Between the States could have ended in 1863, instead of 1865."

"How was that, uncle?" asked Mike, his curiosity stirred.

"Well, one stormy night, the bedraggled Confederate Army, defeated at Gettysburg, and trapped on this side by the rain-swollen Potomac, was allowed by our slowpokey Union forces to escape across the river into Virginia at Falling Waters."

Entering Lock 44, the captain declared "And here we are at Williamsport — the most typical canal town along this 'big ditch.'"

While the *Daisy Mae* crossed the impressive Conococheague Aqueduct, the captain pointed to a grove of trees on the right, surrounding a small house.

"Over there — a stone's throw from this aqueduct and the canal," he said, "is where our friend, 'Hooper' Wolfe, was born on Valentine's Day in 1894." Then he shouted to old John: "Tie up the boat. We're all going into town."

Conococheague Aqueduct

While old John secured the boat to one of the Cushwa Coal Co. wharves, the captain showed Mike where Mosby's Raiders had blown out one end of the aqueduct during the Civil War, in an effort to stop canal traffic.

The Clarks paused in Williamsport to buy fresh new clothing and supplies and some oats as a treat for Bozo and Lazy Bones. Old John was treated to a plug of tobacco. He chewed it most happily with the few remaining teeth he still had in his jaws. As captain led the way out of the tobacco shop, he looked down and tapped the sidewalk with his boot.

"Can someone tell me where this reddish-brown stone was quarried?"

Abby and Mike took only one glance at the worn surface. "Seneca!" they both said.

"Right! Williamsport has several such antique pavements."

On a street corner, the youngsters saw a wood stand with colored streamers wound around its top. A man stood over a large cake of ice, covered with a white cloth. "Snowballs!" he yelled, "Get your snowballs!"

Abby and Mike each handed him a cent from the few pennies of spending money given them by Captain Clark. The snowball man removed the cloth and scraped away on the ice with a bowl-like tool in his hand. Back and forth; back and forth. Then he opened the iron bowl and dumped the ice shavings it held into a thin, splintwood dish.

"What flavor you want?" he said, pointing to the squirt bottles filled with bright red, green, yellow, purple, and orange liquids. He named them so they could make up their minds: "There's strawberry, mint, lemon, blackberry, and orange. Which you want?"

Abby chose mint — it looked so cool. And Mike asked for the blackberry flavor. The man upended each bottle and shook it over each dish of shaved ice, until it turned a deep color. Although they had no spoons, Abby and Mike enjoyed the cool and sugar-sweet, melting, chips of ice that they sipped over the edges of the wooden dishes. But Mike's container leaked and the dark juice ran down his jaw and dribbled all over the front of his shirt.

"You're sloppy," frowned Abby, dabbing most of the stains away with her handkerchief. But Mike just grinned at her and stuck out his tongue at her. And you should have seen how cross-eyed and surprised he was to find that his tongue, too, was a deep purple.

Captain Clark bought himself a new cap; it had a visor of patent leather that glistened almost like a black mirror. Its underside was faced with green. In bright sunlight, the reflection on his large nose made it look like a green cucumber.

Abby chose a similar cap. "But it makes you look even more tomboyish," protested her mother. "So unladylike . . ." But she admitted that the cap held Abby's long pigtails under more control and stopped objecting.

The captain bought his wife a blue sunbonnet. Later, she displayed it proudly for all to see as she stood by the tiller on the stern of the *Daisy Mae,* while Mike rode astride Lazy Bones, with his legs spread wide, admiring the shine on his "cowboy-type" boots.

"That was an interesting old town," said Mike to his uncle at dinner that evening. "Does Williamsport have much of a history?"

The captain's eyebrows shot up. He wiped his lips slowly with a napkin.

"Indeed it does," he said. "Williamsport was famous long ago as an Indian settlement. It was the major supply point of your General Edward Braddock on his ill-fated march to defeat the French at Fort Duquesne in 1755. Braddock's aide was a young officer named George Washington.

"During the Revolutionary War, Williamsport was the capital of the infant United States for 28 hours. In 1790, *President* Washington inspected Williamsport as a possible site for a permanent National Capital.

"His reception was impressive — with great bonfires and brilliant fireworks at each street corner. But Washington decided that Williamsport was too difficult to reach by water travel. Instead, he chose a swampy flat on the Potomac and Anacostia Rivers, near the seaport of Georgetown, as the Federal City of the baby nation.

"Although Williamsport was considered too difficult to reach by water, many small boats were built in the streets of the town named for Revolutionary General Otho Holland Williams. This was long before the canal was even started in Georgetown.

"Slaves used poles to push the Williamsport boats, laden with fur pelts, wheat, grain, flour, eggs (3c a dozen), butter (7c a pound), and lumber, down the Potomac, to the ocean-going ships docked in Georgetown and Alexandria. The trip took one day to streak downriver and through the early canals of the Potowmack Co. organized by George Washington (with steamboat inventor James Rumsey as chief engineer), but five days to struggle back to Williamsport with merchandise.

"That is," said Captain Clark, "*if* the boats weren't broken up and sold in the streets of lumber-scarce Washington City, where wood was needed for new construction."

"And, Abby," the captain slapped the tabletop for attention, "since you like to play baseball with the boys — did I ever tell you that Abner Doubleday, who invented the game at Cooperstown, New York, later became a Union general and defended Williamsport during the Civil War?"

"Really? Well, I'll be jiggered!" exclaimed Abby. And, in her surprise, she almost stabbed her cheek with her fork.

The *Daisy Mae* approached the canal feeder Dam No. 5 on the Potomac — separating Maryland and West Virginia. Groups of men and boys sat on the stone blocks facing the river and the impressive "waterfalls" of the dam. They were fishing for bass and catfish and many strings of flopping catches lay beside them, wrapped in wet burlap.

"A peaceful scene, eh?" remarked Captain Clark to Mike. "Well, sir, it was anything but, when Confederate General 'Stonewall' Jackson attacked that dam from West Virginia in 1861."

"Why did he do that?" Mike wanted to know.

"Oh, he had solid reasons for using solid shot and shell against that dam. He wanted to pound it into smithereens and stop the flow of water into the canal — and with it — all boat traffic.

"You see, the Union Navy needed coal or 'black gold' from West Virginia to fuel its new steam-driven warships on the East Coast. Federal troops needed hay and forage for their cavalry. Everyone needed food. 'Stonewall' soon tumbled to what was going on with all those barges being towed down to Washington and Alexandria.

"From the West Virginia heights overlooking Dam No. 5, hot shot and shell rained down, all day long, on the dam, nearby houses, and any boats seen on the canal. The Federal troops protecting the area used smoothbore muskets and these could fire bullets only a short distance. They couldn't protect themselves, let alone the canal, the boats, and the Baltimore and Ohio Railroad cars laden with coal, a short distance to the north. During the night, however, the Union commander got hold of some sharpshooters with rifles and they soon made the rebels retreat.

"It took a lot of cutdown trees and backbreaking patchwork for the canal crews to repair the breached dam and the banks. But the damages were repaired and canal boats and railroad coalcars soon moved both ways, carrying vital supplies.

"However . . ." Captain Clark shook his head sadly, then went on: "Besides those rebel raiders, there were many Marylanders who sympathized with the Southern cause. Came darkness, they would dig a hole in the bank of a vital stretch of the canal, stick in a charge of black powder and BOOM! BA-R-R-OM! the night sky would light up. Earth, rocks, and towpath would disappear in a black cloud of dirt and water. And so would any canal boat that happened to be nearby. If it wasn't shattered by the explosion, it could be carried, helpless, through the breached hole, atop the water escaping downhill to the Potomac. Yessiree! Our War Between the States was a sad, bad time for our country . . ."

The *Daisy Mae* floated near famous Fort Frederick, built after Gen. Braddock's defeat to protect the colonists against marauding French and Indians.

"George Washington was at Fort Frederick exactly 150 years ago this summer," related Captain Clark. "Later, during the Revolution, the fort was a detention center for Hessian and British prisoners." The captain paused. "Perhaps one of your father's ancestors was among them, Mike," he said, mischievously.

Mike made a wry face — but said nothing.

Crossing the Licking Creek Aqueduct, probably the largest single-arched stone "trough" in America, the *Daisy Mae* approached the ruins of the old village of Millstone Point.

"About a 100 years ago," said Captain Clark, waving his arm to take in the surrounding area, "this was a stagecoach stop on the first Federal road to the West, with a mill and tavern to serve the townspeople and travelers. Later, when the C & O Canal came through, a funny thing happened here."

"What was that, uncle?" asked Mike, eagerly.

"Well," chuckled the captain, "this was long ago. But oldtimers still laugh over the story of a horsedrawn stagecoach, loaded with passengers, stopping at the tavern for the night. But the brakes failed, just as the coach drew up to the inn, and the horses, the stage, and the passengers all landed in the canal. I'm sure those wet and muddy people didn't like their 'baths' *before* supper."

The captain's midriff heaved up and down as he ho-ho-ho'd.

"I don't think that was so funny," commented Mrs. Clark, frowning at her husband. "Think of all those poor women travelers, half-drowned and dragged down by their many petticoats."

"Who said anything about lady passengers on that coach?" demanded the captain. "Besides, if there were any, they probably got the full help and attention of their male fellow travelers. Besides," he winked, "don't you suppose the occasion might have started a romance or two?"

"Oh! You men . . . !" Mrs. Clark was blushing. She stamped her shoe down on the deck and held her fists akimbo on her hips, glaring at her husband. But there was a curl to her lips and she took to hanging her wet laundry a bit faster on the line strung from the cabin roof to the stern water-rail. She hung the clothes even faster, after she turned her head and caught her husband winking at her.

"Well, Mike," continued the captain "as I was telling you, before the interruption — a famous general, the 5th president of the United States, also figured in the early history of Millstone Point. He was Andrew Jackson, hero of the War of 1812. He defeated the British (oops! — pardon me, Mike) at New Orleans and became President in 1829. While visiting hereabouts, a committee from Hancock came out to meet 'Old Hickory.' That was his nickname. Since excavations were going on in the neighborhood, and since black powder was used to blow the rocks apart, the officials arranged to have some of the charges fired to honor the approach of the great man.

"BOOM! BA-ROOM! BOOM! Blasts of flame, smoke, and dirt shot high into the air. One nervous committeeman turned, hearing the scream of an animal, and rushed over to 'Old Hickory,' who sat firmly and calm on his frightened, rearing, horse.

" 'Did the detonations alarm you, general?' " The official was wringing his hands in distress.

" 'No, sir!' came the sharp reply of the warrior-statesman. 'We've heard this sort of "music" before!' "

HANCOCK, MD.

CHAPTER 7

George Washington's "Bathtub."
Danger at the "Devil's Eyebrow."
Indian Fish-Traps.
A Mountain is Bored.

The C & O Canal reached its northernmost point at Hancock, Maryland. There the state is less than 2 miles wide — tucked in between West Virginia and Pennsylvania. Named for John Hancock, first signer of the Declaration of Independence, the town was once a thriving frontier trading post.

Mike had been busy part of that morning spelling old John on the towpath. It was fun driving the mule and watching the sights along the canal: the rabbits, raccoons, squirrels — and now and then a beaver — scurrying to hide as Mike came down the towpath, leading the mule. Mike noted the patterns made by sunlight filtering through the treetops; the branches seemed to wave "Hello" in the breeze as he passed beneath them. He was surprised to see how clearly the berm walls of rock were reflected in the calm canal waters. Like in a mirror. And the graceful bends of the towpath brought into view fresh scenes of quiet and peaceful fields of ripening grain, or meadows of grazing cattle. Later in the day, Mike had finished tidying up the hayhouse, when he saw his uncle standing at the tiller, relaxed and puffing on his pipe, while the *Daisy Mae* was being locked through Locks 51 and 52. Mike went astern.

"Uncle, is there anything interesting roundabout — besides the lovely scenery?"

"Anything interesting about Hancock? I should say! Just give me a moment to collect my thoughts. This warm weather had me day-dreaming." He knocked the dead ashes from the bowl of his pipe against the hull of the *Daisy Mae*, tamped in fresh tobacco, lighted it, and sucked, contentedly, at the stem. A faraway gaze came into his eyes and Mike imagined he could see his uncle's memory at work, swiftly sorting through stacks of mental notes.

"Well, sir," began the captain, watching the smoke rings float apart over his head, "President Thomas Jefferson started the Great National Pike, America's first Federal highway, westward from Cumberland. The road was soon extended eastward and pioneers, headed for the west, poured through Hancock afoot, on horseback, by ox- and horse-drawn wagons, stagecoaches, and carriages.

"By 1817 — and you'll recall, Mike, that was 11 years before the C & O Canal and the B & O Railroad were even begun — the Cumberland Road, or National Pike, was already longer than 100 miles and one traveler reported that 'Old America seems to be breaking up and moving westward.'

"So you see, even though Hancock was already an important frontier town on the National Road, it really began to bustle after the C & O Canal reached there. Taverns and business establishments mushroomed to accommodate the freight wagoners, mail stage drivers, and travelers.

"The destruction of Hancock was threatened in 1862, however, when Confederate General 'Stonewall' Jackson ordered the bombardment of the town, because Union artillery had shelled Shepherdstown, following the Battle of Antietam.

"Well, sir, Jackson demanded that the Union commander of Hancock, surrender the town, or have it destroyed around his ears. The Yankee officer promptly sent word back to 'Stonewall': 'Go ahead! Shell us and be damned! This town is full of Southern sympathizers. The more you kill off, the less for us to worry about.'

"Yessiree! That silenced the Rebel cannon before they even began firing."

Abby appeared suddenly from the cabin, where she had been helping her mother with the "housework."

"Look," she cut in. "Look at that dandy pleasure boat. A steamer, too, flying all those gay pennants — and listen to that music!"

"That's the *Oriole*," said her father. "She's used for picnic and fishing trips up and down the canal. You see, Mike," he glanced at his nephew, "We do have steamboats on the canal. Also, we don't neglect having a little bit of fun now and then along this big ditch."

"Indeed, sir, I can see that," grinned Mike, and his ruddy cheeks grew even redder as he listened to the strumming of the banjos, the rat-a-tat-tat of a drum, and the blare of trumpets, that accompanied the laughter coming from the passing *Oriole*.

"That music was fun!" Abby was sparkling and tapping her toes. Even her mother smiled and nodded her head, as the pleasure craft disappeared downstream.

"And will you look at all those apple and peach orchards on both sides of us." Abby had already turned her attention elsewhere. "I wish they were ready to eat. Then we could stop and pick a few."

"*Buy a few . . .*" corrected her father, sternly, and his bushy eyebrows moved into a stern line. He slapped his hand down on the tiller, emphasizing each word.

The *Daisy Mae* passed a mill and Captain Clark pointed to it with his pipe, telling Mike that water from the canal was purchased by the mill owner to operate a waterwheel. "It grinds up the leaves from sumac trees and shrubs that grow around here."

"Whatever for?" Mike was puzzled.

"Glad you asked that question," said his uncle, around the stem of his pipe. "The extract is used for dyeing or tanning leather hides. You know, for making into shoes, boots, luggage, and such items . . ."

"As trouser belts," cut in Abby. "Father sometimes threatens to spank me — 'tan my hide' he says — with his belt, when I'm naughty. But he never does," she added, grasping her parent's hand and flashing an impish grin up at him.

The captain wagged a warning finger over her. "Some day," he shook his head, slowly, "some day . . ." He then turned to his nephew.

"Six miles up that road, Mike, is a famous old watering place. Your father would call it a health spa or 'bawth.' Hundreds of years ago, its mineral springs were discovered to have great healing powers. Indians drank its waters to cure their ailments and bathed the wounds of their warriors in the shallow, rock-bordered pools. Your English Lord Fairfax named the place Bath, after a resort in the Motherland, and gave it to the colonists before 1776.

"Even George Washington's 'bathtub' — a hollowed-out rock in the springs — is located there. In fact, George and his two brothers often visited Bath. George bought some land there and built himself a house, a kitchen, and stables. The gay resort often welcomed other personalities of the time, including Thomas Jefferson and the French hero, Marquis de LaFayette.

"The city's now known as Berkeley Springs," said the captain. "It's in West Virginia and boasts a mysterious old stone castle, high on a ridge, over the springs."

"A real castle — like those in England?" Mike wanted to know.

"Like in England," nodded his uncle. "And complete with strange tales, ghosts, and secret passages."

Later, about an hour's tow up the canal, the *Daisy Mae* approached the "Devil's Eyebrow," a large, beautiful arch on the earth's surface. High above the berm side of the canal, it formed a cave shelter in the shape of an eyebrow.

Captain Clark made a suggestion. "Suppose I call old John to rest with Bozo on the towpath, while you and Abby explore the cave up there?"

"Yea! That's peachy," said Abby.

"A bit of all right," agreed Mike.

"But remember," warned the captain, "don't be gone longer than a few minutes, else we may get in the way of a passing barge." Noting that the youngsters were already removing their shoes and outer clothing, he steered the boat close to shore. Abby and Mike dove overboard, swam a few yards, then scrambled up the berm. Turning, they waved toward the boat, then ran up a tree-covered path leading to the cave.

"Come on; hurry, you slow poke," cried Abby over her shoulder. "We've got only a few minutes. You heard father."

"But I'm . . . I'm out of breath," gasped Mike.

Abby came back and, placing her hands behind her cousin's shoulders, shoved him up the incline. They reached the cave and began to enter it — with Abby still pushing Mike ahead — when he screamed:

"Snake! . . . Snake!" His eyes were wide with fear and he pointed into the shadows at his bare feet.

Abby's hands flew instantly over her cousin's shoulders. She flung him sideways — into the bushes near the entrance.

Then she looked down at the large reptile wriggling and disappearing into a hole beneath a rock.

"You sissy!" She turned on Mike. He lay trembling and sobbing in the bushes. "'Fraidy cat! That was only a harmless blacksnake."

But when she saw that Mike's face was pale and twisted and that his teeth chattered, she quickly bent and pulled the boy to his feet.

"Sorry, cousin, I'm sorry." And she kissed his tear-wet cheek, trying to comfort him. "Come on — let's get back to the boat."

She crooked her arm around Mike's neck and shoulders and helped him down to the canal.

"Back so soon?" called Captain Clark, at the tiller. "What's wrong?" He could see Mike was sobbing as Abby led him into the water. But she did not reply.

The swim back to the boat calmed Mike, and when Abby shoved him up and over the side, and climbed aboard herself, Mrs. Clark was ready with towels to help dry the youngsters. She was shocked with alarm when Mike blurted: "We found a snake in the cave. It was about 10 feet long," he said, trying bravely to control himself, "and Abby rescued me from it."

"Oh, poof!" snorted Abby, answering the questions in her parents' eyes. "It was only a blacksnake scuttling to get away from us. And it wasn't more'n 3 feet long. But we know now, I guess, why that old cave up there is called the 'Devil's Eyebrow.' " She patted her cousin's arm, "Don't we, Mike?"

About an hour after passing Leopard's Mill (Captain Clark patiently explained to his nephew that it was named for its human owner and was not a place where leopards mill around), the *Daisy Mae* approached Dam No. 6.

"That poor old dam," said the captain, shaking his head, "has had its share of troubles. Floods, many of them; then a bad fire. It was the only dam to ever burn along the canal.

"You see, some fishermen built a fire to warm themselves at low tide — over the logs of the dam. The wood structure caught fire and was destroyed — spilling out all the loose rocks penned up by the wood cribbing. Oddly enough, this dam had survived the Civil War. Cannon fire from Confederate artillery couldn't blow it apart.

"And, Mike — want to hear something about the redskins who lived around here?"

"My word! They were everywhere, weren't they?"

"Just about. Especially up there ahead, south of Little Orleans. Some rock ledges stretch out, like arms, into the river. The redskins used these extended rock bars to trap fish for their meals, long before the first white man appeared here.

"And, in 1760, a new road passed through Little Orleans, to link Fort Frederick with Fort Cumberland. For many years, traffic of all kinds used the route: men on horseback, supply wagons, and stagecoaches — until a more direct highway was opened through the mountain passes."

The *Daisy Mae* rose higher and higher through a succession of locks. Out of Lock 66, it was being towed toward the Paw Paw Tunnel, when Abby tugged at her cousin's arms.

"Up there, Mike, is the 'leaking water spigot' I told you about. I remember it from last year. Mother had me bring down two buckets of spring water. And did it ever taste fresh and cool and good to drink!"

Slowly, the *Daisy Mae* was drawn from the bright, warm sunlight into the cool shadow under the edge of a cliff. It loomed over the deckhouse roofs. Covered with moss, the rock overhang shed water in rivulets and droplets over the boat, splattering over the children and Captain Clark.

"Ooo-h-h!" gasped Mrs. Clark and scooted for shelter in the cabin. A stream of cold water had dribbled down the back of her neck.

Now the steps leading to Abby's "water spigot" were visible; boatmen used them to reach the stone, wood-roofed, springhouse atop the cliff.

"Sorry we can't stop there now," said Captain Clark. "Perhaps on our return trip . . . I'm told that the cliffside becomes a magic mountain of ice in wintertime — when the canal is drained and isn't used. But the freezing springwater keeps oozing down and over that rock overhang. The massive ice growths must look very interesting — especially when they sparkle in the sunlight."

"Do icicles also form there?" asked Mike.

"Giant ones, no doubt," nodded his uncle. "Now, up ahead, near the tunnel's north portal, you'll note similar rock cliffs overhanging the berm. Sometimes, massive slides of rock come tumbling down, filling the canal bed and stopping navigation."

Abby and Mike shuddered.

"But you needn't be afraid," the captain raised his hand, reassuring them. "When cracks are spotted by the work crews, they warn the boatmen to steer clear of the possible slides.

"Be sure you look up when we come down along the steep gorge at the north portal. Over that entrance to the tunnel you'll spot a fold in the ancient rock. Formed by some earthquake — millions of years ago. That natural arch was counted on by the clever tunnel builders, because they realized it would strengthen the top of the tunnel and prevent rock falls inside the passageway during the construction work.

"Sheets of water often stream down those cliffs over the portal — coming from springs and runoff — just like Abby's 'leaking water spigot.' And here, too, I suppose, frozen waterfalls and giant icicles hang down over the cliff. Must be a glittery wonderland in wintertime."

The Paw Paw Tunnel, the captain went on to explain, chiefly for Mike's benefit (Mrs. Clark and Abby had already heard the facts), was one of the canal's most impressive features. It was built to bypass some very difficult places along the Potomac River in the Paw Paw Bends. The river looped through 6 miles of hilly country, so the tunnel was designed to cut across one double loop with a tunnel and save the building of many locks. The tunnel was bored through 3117 feet of mountain rock (you can walk through in 20 minutes — or 10 minutes if you're scared of dark places). It became one of the "wonders of the world."

Laborers were imported from Ireland, masons from England, miners from Wales. Also local Pennsylvania and Maryland masons and laborers. They were called "Pennsylvania Dutchmen."

The boss was a tough Methodist minister. He used kind words, harsh commands — or an ax handle over the head, if necessary, to get the work done, or to settle arguments among the quarrelsome Irish and German workers. It took 14 years to finish the greatest engineering project along the entire canal, in spite of all back-breaking, heart-breaking troubles.

Giant boulders in the cliffsides, related Captain Clark, had to be blasted apart with black powder. Men then used sledges and picks to break the large pieces of rock into smaller, portable, chunks. This rubble was carted away to miniature, manmade mountains by mule-drawn railcars.

The captain pointed to several such "spoil heaps" as the *Daisy Mae* was towed nearer and nearer to the north portal. Then he chuckled: "Just recently, I heard a funny story about the tunnel. Want to hear it?"

"Of course! Of course!" yelled Abby and Mike.

"Well, it seems that this Irishman operated a sort of elevator (you'd call it a 'lift,' Mike) at one of the vertical shafts. It had been driven deep, down through the cliff, so air could get into the tunnel. The Irishman's job was to lower men and supplies into the shaft and bring up to the surface the loads of rock that had been blasted and chopped apart.

"The Irishman used a mule to pull the elevator up or down by pulleys and ropes. And the man was always cursing his beast and the mule was always stubborn and mean and slow. But the work somehow went on; the spoil heaps grew higher; tempers flared up, then simmered down.

"One twilight, however, the mule balked and wouldn't lift some laborers to the surface at the end of a hard day's work. The Irishman flew into a rage and whipped the poor animal until it finally moved, bringing the workers to the surface. Then, the Irishman happened to turn his back for a moment. In a flash, the mule reared both hind legs and kicked its master full in the seat of his pants.

"That did it! The man yelped with pain, then whirled around and kicked back at his mule."

Everyone was laughing so hard at this point, even the captain. He couldn't continue his story for some moments. Finally, he drew his handkerchief, wiped his eyes, blew his nose, and went on:

"Unluckily, the mule had been standing at the edge of the shaft when it was kicked. Down it fell into the deep hole, scraping its hide, but landing alive at the bottom. It started squealing and squalling. Since there was then no way to get the beast back to the surface, its master, in addition to his other duties, had to lower bales of hay and buckets of water down the shaft to keep his mule alive. Until workers could link up the vertical shaft with the tunnel diggings nearby.

"Now that story is just a wee bit hard to believe," ended the captain, "because the air shaft is several hundred feet deep. But there is a believable tale (I think) about what once happened in that same Paw Paw Tunnel.

"But I don't suppose you want to hear this yarn," he teased, turning to shade his eyes and peer up the canal at the portal, now partly obscured by several boats ahead of the *Daisy Mae*.

"Yes, we do! Come now, uncle, tell us." Mike had overcome his usual shyness and his voice was shrill and insistent. He even tugged on the captain's sleeve, until the boatman turned with a smile to tell this story:

"Well, Mike — in the old days it was the custom for a towboat captain entering the tunnel at night to send a boy or a muledriver like you ahead to post a lantern at the far end. That light would warn oncoming boats that the tunnel was already occupied. It was a sort of would-you-please-wait-your-turn? procedure.

"This didn't always work. Sometimes, a roughneck captain would order his canal boat to be towed in, regardless of the warning lantern. So the mules would meet, head-to-head, on the towpath in the middle of the tunnel and the stubborn captains would yell and scream at each other to get out of the way.

"Once, neither of the opposing captains would give in. And soon — bitter words, then fists, flew back and forth between the battling crews. This went on for a day or so, while boats piled up for miles at each end of the tunnel. Bets were made on who'd win the argument. The canal company officials were furious, but they couldn't get the quarreling captains to move out of the tunnel.

"At last, the section superintendent became desperate. He rode his horse to a nearby farm and bought all the green cornstalks he could find. Tying them to his saddle, he rode swiftly to the upwind end of the tunnel. There he built a roaring fire and into the flames he tossed his green cornstalks. The choking smoke — lots of it — went billowing into the tunnel.

"Very soon, the pentup boats and their crews — red-eyed and weeping from the smoke — came streaking out into the fresh air at top speed. You never heard such cussing, and coughing, and choking sounds," grinned Captain Clark, ". . . and that's no choke," he ended with a pun that made his wife wince.

"Sir!" She had a glint in her eye. "Soon you'll be changing the name of that tunnel for the sake of a pun — so the children will laugh — from Paw Paw to Haw Haw."

The captain's jaw dropped in mock dismay. " 'Haw Haw' . . . I give up," he moaned. He threw up his arms in surrender and winked at Abby and Mike. But they kept their faces frozen and stared into space, pretending they did not see him, though they had trouble controlling the laughter welling up in their throats.

DAISY MAE

CHAPTER 8

Out of the Tunnel, Into More History and Adventure. Black-Eyed and Blue-Eyed Twiggs. Old-Time Oldtown. The Daisy Mae *is Lifted Higher Than the Washington Monument.*

Light was fading when old John led Bozo, with Mike on his back, into the dark shadows of the Paw Paw Tunnel's north portal. The captain had lighted the large headlamp fixed to the top center of the stable. The shiny reflector sent warm, golden beams of light stabbing through the ghostly, white vapors that floated up from the black surface of the water.

John swung his lighted lantern. Mike listened to the clippedy-clop! clippededy-clip! of hooves on the towpath. A heavy, wooden handrail prevented mules and hoggees from falling into the dark canal. The shouts of the muledrivers bounced off the brick walls and ceiling of the tunnel, which was shaped like an upside down "U."

Mike wasn't really afraid of the flickering shadows. But they *were* a bit spooky, so he whistled to keep up his courage, while watching the tiny spot of sunset at the tunnel's far end. It seemed that he would never reach there. Beads of water dripped from the brick ceiling; the tunnel smelled of mold. Then Mike became fascinated by the wriggling lights of the lanterns reflected by the water in the canal onto the curved ceiling. As the tunnel's far end grew larger and brighter, Mike grew more cheerful.

Suddenly, he turned on Bozo's back and shouted at the shadowy boat, with the lamplight pouring from its windows. "Ahoy, back there. Ahoy, Uncle!"

And Captain Clark replied "Ahoy, up there. Ahoy, Mike!"

Now the lad really felt better, hearing familiar voices bouncing all around him. He slapped his new boots against the barrel-shaped sides of Bozo and listened to the rhythmic clatter of the mule hooves and the echoes. Soon, the *Daisy Mae* was outside, in twilight, and away from the tunnel's gloom. Mike slid off Bozo's back and returned to the boat.

As the boat was towed up-canal, Abby looked back and indicated the stone steps on each side of the south portal.

"I once climbed them with father," she told Mike, "right to the very top. And on the berm side, there, we found some fossil shells in the shale. They prove, father said, just as he told us at Great Falls, remember?, that all of the Potomac region was once deep underwater."

"Do you have any of those fossils?" asked Mike.

"Oh, yes. Father picked out some pieces of shale with a knife blade. I have them in my collection box."

"Will you show them to me?"

"Sure; later. I may even let you have one or two pieces to show your schoolmates back in England."

"I say, cousin, that's mighty good of you. Gee whiz!" he ended, excitedly, slipping into American slang.

The next morning, the cousins awoke to find the canal curving ahead, serenely, where the towpath was flanked by trees. Wildflowers grew along the wet banks on both sides of the canal: strawflowers, blueweed, daisies, wild roses, Queen Anne's lace, butter-and-eggs, and leather-flowers, known also as vase vines. All the blackberry and blueberry bushes had been stripped of their delicious fruit by birds and the families of passing boatmen. Some Jacks-in-the-pulpit and Dutchman's breeches lingered over from Spring and the youngsters were delighted with their beauty.

The early July air was warm; humid, in fact, but not too uncomfortable, and Mike, especially, was pleased at the floral display that reminded him of the gardens in England, where he liked to roam with his parents.

He felt homesick for a moment, then . . .

"Look there," he cried. "What are those big, strange birds lurking near the woods and going 'Gobble! Gobble! Gobble!' "

"Silly . . ." giggled Abby. "Those are wild turkeys. And they're good to eat, too, especially when mother roasts them with chestnuts at Thanksgiving."

Then Abby recited all she could remember from her classroom studies about the Pilgrims.

"They left Plymouth, England, on the *Mayflower* and started the Plymouth Colony in the New World. And some friendly Indians — I mean, redskins — showed them how to plant corn. So when they had a good harvest the next year, the colonists shot some wild turkeys in the woods and roasted them for a big Thanksgiving dinner."

"With cranberry sauce," added Mike, proudly, remembering now his mother's description of the Thanksgiving dinners she enjoyed, before migrating to England.

"No, silly. I don't think the Pilgrims ate cranberry sauce with their turkeys," said Abby. Then she hesitated and scratched her forehead, thoughtfully: "But my schoolteacher — she once taught in Boston, near Plymouth — did say that lots and lots of cranberries grew around there . . ."

That fine point could not even be settled by Captain Clark, when he was questioned.

"But since you're interested in redskins just now," he said, "we're passing through an area occupied by them thousands of years ago, judging from the large numbers of spear points, arrowheads, and worked quartz rock, which boatmen and their families have found hereabouts." He paused to signal old John and Bozo to go faster. "We're now about 25 miles from Cumberland," he announced.

Passing through Darkey's Lock (No. 67), the captain observed that this had been a landmark for boisterous canallers. "Those oldtimers," he said, "had a gay time whistling at tavernkeeper Gene Stumph's seven beautiful red-haired daughters." And Mrs. Clark, hearing this, glanced sharply at her husband, as though she suspected *he* might have been one of the oldtime whistlers.

"We should be entering Lock 69 about noon," said the captain. "That's Twiggs Lock. Mike — do you know what a feud is?"

"I know! I know!" shrilled Abby, jumping up off the deck where she was sitting.

"I didn't ask *you,*" said her father, pointedly, like a schoolmaster. "But since you've interrupted, suppose you spell the word and give its meaning."

"F-u-e-d," spelled Abby, promptly, "and it means family fights."

"Your spelling skills need more attention, young lady," said her father. "It's f-e-u-d. But you're correct about the meaning: a family quarrel or fight. Many ended most violently.

"Anyway, the Twigg family was always active along this canal section between Little Orleans and Cumberland. The first Twiggs, John and Rebecca, built a cabin near here at Sink Hole Bottom, in the mid-1700's. They had two sons: Robert and Fleetwood John. When he grew up, Fleetwood John spent much of his time on hunting and fishing trips into the woods. Once he was gone for several weeks. Then he turned up at the family homestead with an Indian maiden trailing behind him.

"Old John Twigg refused to have his son and the redskin girl live under the same roof with his family. So Fleetwood and his new wife went off and built themselves a cabin at the edge of a meadow, separated by a large pond from his father's house. There, Fleetwood and his squaw started raising black-eyed little Twiggs.

"Later, brother Robert got married and began rearing blue-eyed little Twiggs on his side of the pond. But his blue-eyed children were never allowed to speak to or play with their black-eyed cousins.

"This situation continued for about 30 or 40 years, until a blue-eyed Twigg fell in love with a black-eyed cousin. That's when the Twigg feud ended happily."

"I like that story, uncle," said Mike, crinkling his dark eyes in a smile. "And although I'm younger than Cousin Abigail, she does have blue eyes and perhaps I'll marry her some day."

"Not a chance!" exploded Abby, frowning.

"Children!" Mrs. Clark was wagging her finger. "Now don't be starting a feud of your own. You can settle that question many years from now. Meanwhile, here are two baskets. Father can let you off at Lock 70. That's at Oldtown. Some wild elderberry, raspberry, and strawberry bushes thereabouts may be ripe for picking. Bring back all the berries you can — but be careful about the poison ivy and oak. You know what they look like, Abby — so stay away from them."

The youngsters leaped ashore at Lock 70, where a wooden covered bridge, the lockhouse, and store had burned down only 2 years earlier.

Abby and Mike found a hillside near the lock covered with flowers. Various birds also seemed to like the area. Although they could not identify them all, Abby and Mike did see some quail, mourning doves, bluebirds, cardinals, chickadees, flickers, and other woodpecker species. Of course, wherever there were small birds, the larger, predatory birds would also go. Abby spotted two hawks circling overhead. And once, coming down a path near some woods, the two cousins were startled by a sudden loud slap! slap! of wings. A turkey vulture rose awkwardly into the air from bushes just in front of them. A rabbit it had caught, and had been devouring, hung limply in the bird's beak.

"Ye-eech!" cried Mike and hurried on, almost stepping on a tree frog he hadn't seen too well, because of its camouflage coloring.

The children found some berry bushes, but the berries were too small to pick. "Mother won't be pleased," said Abby, "but I'm sure she'll believe we found none worth picking."

Mike appeared anxious. "She may think we returned with two empty baskets because we ate the berries on the way back."

"No," replied Abby, "I think mother just wanted us to get some exercise — away from the boat. After awhile, you know, you get to feel mighty cramped on the *Daisy Mae*."

She turned back on a trail, angling away toward Lock 71, upstream. There they could board the boat.

Suddenly, Mike screamed: "Snake! Snake! There's a big snake. See — there to the right, near the rock."

"Not again!" blurted Abby. "You certainly are a scaredy cat when it comes to snakes." She stepped over to take a closer look at the slowly curving coils of the reptile.

"Good gosh!" she cried, her eyes widened. "Let's get away from here. That's a poisonous copperhead!"

As she raced past Mike, she grabbed his hand, jerking him along. They both went running down the path to the canal lock.

Mrs. Clark was glad to forgive them the empty berry baskets, when she heard about the copperhead.

"Land sakes alive! You have to keep a mighty sharp lookout for snakes everywhere in the wilds, especially near the canal. But don't you worry about the berries; we'll just have chopped-up apples and raisins for dessert tonight."

When he learned of the youngsters' latest escape from danger, Captain Clark also reminded them that wild areas anywhere were dangerous for every living thing — bird, beast, or man. Each tried to conquer the other, take its life for food, or in fear, or to acquire new territory.

The *Daisy Mae* was approaching Oldtown and the captain commented that many interesting things had happened there.

"Want to hear them, Mike? That is, if you've quieted down from your latest snake scare?"

"Yes, sir. If you please. I feel quite tip-top now, thank you."

"Good! Well, long before any white man appeared hereabouts, redskin tribes met and traded along the famed Warrior's Path. It forded the Potomac — at Oldtown. Four other trails, besides the Warrior's, led into Oldtown across several mountain ridges. We call them Bear Hill Trail, Creek Road, Dry Run, and Mill or Big Spring Trail.

"In 1692, a number of Shawnees settled there. Fifty years later, the redskins were gone and the first white settler came. He was the celebrated Colonel Thomas Cresap, a well-known pioneer, patriot, and businessman in Colonial days. Like yourself, Mike, he was born in England. But about 200 years earlier, in a town called Skipton — not far from your own birthplace.

"Cresap came to America when he was twice your age, Mike. He got married, and one of his three sons (he also had two daughters), was named Michael — just like you. The colonel rented a large farm on the lower Potomac near what is now Mount Vernon. He ran a ferry, got involved in various border conflicts, and also acquired a house on Antietam Creek, near the Potomac.

"Then he bought 200 acres at Shawnee Old Town and built a stockade fort. He named it Ft. Skipton, after his birthplace. He planned the first town in Allegheny County and built a trading post.

"Well, sir, Col. Cresap's fortified house became a true historical landmark. When George Washington was 16 and working as a surveyor for Lord Fairfax, he stopped for some time at Ft. Skipton. Later, he was there with Gen. Braddock and the troops fighting the French and Indians. And in 1775, Washington and other revolutionaries, inflamed by your King George III, met in Michael Cresap's 'House of History' — it's still standing on Main Street — before marching to Frederick, then on to Boston — where Washington became Commander-in-Chief of the Continental Army."

The captain paused to gaze directly at his nephew. "I hope my sister taught you, Mike, that George Washington came from an old English family in Northamptonshire — in the center of England. His ancestors figured in English history after the 12th century. So you have something in common with 'The Father of Our Country.'

"Getting back to Cresap's Old Town in the late 18th century — you can picture the hundreds of pioneers passing afoot, on horseback, by carriage, or in wagons, over the main street, enroute to Ft. Cumberland and the land west of the Allegheny Mountains. Oldtown was a busy stopoff for westbound stagecoach lines until a shorter route was built across the mountain ridges. It became even busier, of course, when this C & O Canal was opened.

"And . . ." the captain turned to Abby, "about where you and Mike came flying aboard at Lock 71, in a field overlooking the Potomac, stand the ruins of a stone chimney of the fortified house that Col. Thomas Cresap built in 1742."

The *Daisy Mae* was halted briefly for fresh water at Blue Spring, one of the largest springs in the Eastern United States.

"Doesn't this cold water taste good?" remarked Abby, sinking the dipper again into the pail.

"Um-mm, indeed it does," agreed Mike. "Just like icewater." (And it did, for the springwater temperature averaged 54° in summer.)

The *Daisy Mae* passed through Lock 72 and "The Narrows," a tight place between the river and the ends of Nicholas Ridge and Irons Mountain. It "stepped up," uneventfully, through Locks 73, 74 — and then — 75, the last lock of the Chesapeake and Ohio Canal.

But 9 more miles of level travel lay ahead.

"Actually," Captain Clark told Mike, "only 74 locks were built. But figure an average height of about 8 feet for each step-up. Then consider that we've been lifted into the Allegheny Mountains by those 'steps.'

"Do you realize that the *Daisy Mae* — and we — are now 600 feet higher than Georgetown? We're much higher than even the Washington Monument!"

"By Jimminee!" Abby said, passing by and catching the end of her father's remark. She clapped a hand dramatically to her forehead and pretended to stagger. "No wonder I feel dizzy at this height."

CUMBERLAND
Cumberland . . .

CHAPTER 9

Journey's End: Cumberland! George Washington and the Queen City of the Alleghenies

Abby and Mike were sprawled on the hatch covers, studying a sheet of paper on which they had written some of the odd names of places passed along the canal. Scribbled place-names like Horsepen Branch, Bald Eagle Island, Chisel Branch, Turtle Run, Indian Flats, Calico Rocks, Huckleberry Hill, Horseshoe Bend, Devil's Alley, and Pigman's Ferry.

"Don't these names make you giggle?" asked Abby.

"Oh, grown boys don't giggle," replied Mike, sternly. "Besides, back home in England, we also use queer names for places."

Just then he looked up, startled by a large packet boat, towed by fast horses — not mules — bearing down on them from Cumberland. These swift passenger boats of the mountains always had the right of-way over barges.

Captain Clark cupped his hands around his mouth and yelled: "Lower the towline!"

Old John was already unhitching the rope from Lazy Bones' collar: he knew what to do. But he turned around, hearing the captain's shout, and the rope fell from his hands to the towpath, twisted over the bank, and dribbled off, underwater. The packet swooped over it and passed the *Daisy Mae.*

Abby and Mike watched the steersman reach for his long, straight, tin bugle. Mike thought it resembled an English fishmonger's horn. The steersman put it to his lips and blew three sharp notes of "Red Rover." It sounded like "Lock Ready! Lock Ready! Lock Ready!"

And the nervous lockkeeper and his family streamed from the lockhouse of No. 75, far back of the *Daisy Mae,* then scattered to man the gates. Everyone was scurrying to lock the noisy passenger boat through and away, without a moment's delay.

Captain Clark looked back and remarked, a bit sourly, that peddlers, politicians, and medicine shows, in their hurry to part people from their money, were more apt to travel aboard packet boats that honest citizens usually avoided.

"Shall I fish up the towline, captain?" offered Mike, stomping the heels of his boots down on the deck and standing stiffly in a mock British military salute.

"That would be fine, mate," responded his uncle, returning the salute, while keeping a solemn face.

Mike kicked off his boots and dove into the cool, refreshing canal.

In minutes, he had the soggy rope safely in old John's fist. The chore was usually handled by dogs kept by some captains on their barges. The animals would have the line tied to their collars, swimming to the towpath as Mike had done.

"But you're much smarter, Mike, than any trained dog," teased old John, patting the boy's wet head. "Here is a bright, new cent for your trouble." And Mike was pleased with the copper Indianhead coin. He hitched the wet line back on the collar of Lazy Bones.

Captain Clark steered the *Daisy Mae* close to the towpath, so Abby could jump ashore with some dry clothes for Mike. Besides, the youngsters had permission to help drive the mule on the last lap of the canal. Old John came aboard to take a snooze in the hayhouse. There would be much work for him when Cumberland was reached.

"On shore, there," called Mrs. Clark, standing at the bow. She drew back her arm and tossed a cloth bag through the air. Abby took one step and the bag sailed neatly into her hands.

"For you and Mike," said Mrs. Clark. "Some cookies I've just baked."

"Yummy," said Abby, already nibbling on one. "And are they hot! Here's your share, Mike."

Mike was behind some bushes, changing into the freshly laundered clothes. He came out, running to catch up with Abby and the mule, while buttoning up his shirt. He gobbled down several of the warm cookies. Then he held Lazy Bones' bridle, while Abby swung onto the mule's back. The captain steered away from the towpath.

"On to Cumberland!" he cried.

Lazy Bones moved up the towpath, straining into his collar. He seemed aware that the end of the journey was just ahead.

Suddenly, a group of young people on bicycles came swirling down the towpath, followed by a cloud of dust.

Lazy Bones was startled. Although Mike had a firm grip on his bridle, the mule stopped abruptly, brayed, then swung his body from side to side. He was confused whether to move to the right or left. Abby lost her balance and fell off his swaying back.

"I say, chaps! Watch where you're going," shouted Mike at the passing cyclists. But they stuck out their tongues and raced past.

"Go fly a kite!" jeered one, over her shoulder.

"Go chase yourself!" called her companion.

"Shoo fly — don't bother me!" yelled a large boy, standing up and churning at the footpedals. His bicycle whirled down the path and was gone.

"Some city slickers with bad manners," snorted Abby.

"Should have their mouths washed out with soap," agreed Mike. "And I could do with an overall bath, myself." He slapped away at his freshly laundered clothes — and almost vanished into a cloud of brown particles. "Whew!" he gasped. "All that dust they stirred up."

With Abby riding and Mike leading, Lazy Bones pulled the *Daisy Mae* over Evitt's Creek Aqueduct, then over the city limits of Cumberland. Past low cliffs with interesting rock formations; past woodlands.

The boat bore away from the railroad tracks and was towed near a field of dandelions. Several goldfinches flitted back and forth over it, like bright flashes of sunshine. The boat veered right, into the city's factory district. The towpath ran along a high embankment which sloped left to the river. And from this vantage point, Abby and Mike admired the splendid view of the old city.

They could see the streets, thronged with busy people; the shops, with awnings reaching across the sidewalks; the chuff-chuffing locomotive in the B & O Railroad yards, crested with plumes of black smoke and white steam. The buildings and church spires, clustering the hills, looked very picturesque, in the fading sunshine. Especially to Mike. He smiled strangely up at Abby, astride the mule.

"That scene reminds me," he said, softly, "of some towns near my home in England."

Abby sensed his homesickness and slid off the mule's back to walk the towpath with Mike. They trudged off, walking on either side of Lazy Bones, leading him into the busy Cumberland Basin.

Two arms of this Basin stretched ahead. One led straight to the guard lock and inlet behind Dam Number 8; the other formed a huge pool of water, reaching out to the right and extending well into Cumberland's business district. Close by and to the right, a long strip of land jutted out into the Basin, carrying railroad tracks to boat-loading stations that circled the sides. Here lay canal barges, with their hatch covers removed. Coal cars, positioned overhead, dropped their "black gold" contents through their bottoms — straight down into the boat holds.

Beyond the Basin were streets crowded with stores and saloons that catered to the canallers. Many spent the winters there, aboard their boats, while the canal was closed to traffic. Of course, they had to stock up food in advance, because many places, and especially the lockhouses, could be reached only on foot in wintertime. So they laid in supplies of flour, cornmeal, buckwheat, sugar, molasses, dried fruits, and other staples.

Several large shops nearby were busy building canal boats, and a near-complete barge rested on a sloped ramp, almost ready to launch.

A canal company steam dredge and several repair scows were moored near a lumberyard, where Captain Clark expertly guided the *Daisy Mae* into a berth.

A slight bump on the bow fenders — and the boat came to rest. Old John and the captain made fast the lines around a snubbing post ashore and the ring bolts in the *Daisy Mae's* hull. Her voyage up-canal was ended.

"Welcome to Cumberland — Queen City of the Alleghenies," sang out Captain Clark, as Mike and Abby came aboard, leading Lazy Bones across the fall-board and into the stable. The captain beamed, removed his peaked cap and wagged it in the air. Mrs. Clark also was smiling: because the journey had ended . . .? Or because she was thinking of all the interesting shops and places to visit in town, while the boat was being unloaded and made ready for the return trip . . .?

Mike was fascinated by all the activity thereabouts — especially around the C & O Canal Company's boatyard. This was a 2-story building, with steps that led to the towpath and loading yard.

The boatyard included a blacksmith shop. Here a young man cranked the handle of a geared blower; it forced blasts of air into a fiery bed of coke where iron rods lay, glowing white-hot. Next to this furnace, or forge, on the dirt floor, stood part of a tree trunk. An anvil was secured to its top. It was manned by a giant of a blacksmith. He wore a leather apron and he pounded, with his heavy hammer, on a piece of orange-red strap iron grasped by the jaws of tongs in his hairy fist. The bell-like clangedy-clang-clang! Clang! Clang! rang out like musical sounds. Orange-yellow-white sparks showered all about the grimey, sweaty, blacksmith.

Mike's gaze passed over the carpentry shop; over the tool and parts shop; over the littered work crew equipment, and came to rest on a large outdoor fireplace that supported a huge, black kettle.

He looked up at his uncle, with a question in his dark eyes. The captain had already caught the direction of his nephew's curiosity.

"That isn't a soup kettle," he chuckled, "nor was it used by cannibals to cook missionary stews. Pitch is heated in that kettle. When the pitch becomes syrupy, long, ropelike strips of hemp are soaked in it to become 'oakum.' These strips are packed or 'caulked' between the wood seams of boats to make them watertight. We'll have to have some of that repair work done also on the *Daisy Mae*, I expect."

Captain Clark excused himself, wrote a note, and asked old John to deliver it to the consignees of his cargo. Then the captain returned to his nephew.

"Looks somewhat quiet around here just now," he said, "but there was a lot of excitement in Cumberland, almost 58 years ago, when the entire canal was declared open, from Alexandria, Virginia, Washington City, and Georgetown, clear all the way up to this spot.

"Bands played, guns fired salutes, officials of the C & O Canal Company made speeches to distinguished guests from Maryland, Virginia, and the District of Columbia. Soldiers in colorful uniforms, and cheering citizens paraded in the streets. Canal boats, decorated with bunting, and crowded with passengers, moved down the canal. One boat was constructed of timbers which had been trees, growing atop Haystack Mountain, only 5 weeks before. Now it was loaded with coal and a merry crowd, underway to tidewater at Georgetown.

"But it took 22 years and $22,000,000 (or $120,000 for every mile of that canal), for that event to take place at Cumberland on October 10, 1850. Twenty-two years, plus the heart – and back-breaking labor of thousands of men — many of whom perished along the canal route, together with their exhausted mules."

The captain pictured for Mike the Cumberland of the redskins and the early white settlers.

"The Shawnee Indians had their villages of wigwams all throughout that lowland, when white trappers, hunters, and pioneers began appearing here. A storehouse and trading post was constructed of logs in 1750. In 1753, young George Washington was an adjutant-general of Virginia Militia. He came through here on his mission westward, carrying Governor Dinwiddie's message from Williamsburg, warning the French to keep out of British-held lands.

"You see, Mike, Washington was still an English subject, like yourself. As a British Army officer, he later became active around here, fighting the marauding Indians led by the French. He helped build Fort Cumberland. It stood on that high hill," and the captain pointed north of the canal.

"See that steeple? It's atop the Emmanuel Episcopal Church that now stands on Washington Street, over the site of Fort Cumberland. But, beneath that church are Sunday School classrooms and the church library. These were converted — would you believe it, Mike? — from storerooms for gunpowder, shot, and shell.

"The fort's defenders also used underground tunnels to get water from Wills Creek. And if it became necessary, those same tunnels could be used to escape from the enemy. Many times, the fortified stockades held hundreds of frightened settlers. They were anxious not to have their scalps — or those of their wives and children — get into the hands of the bloodthirsty savages or their French allies.

"But in 1755," continued the captain, "General Edward Braddock arrived from England. With Washington as his aide, he marched 2,000 soldiers openly through Cumberland and into the wilderness. He wanted to push the French and Indians from their stronghold at Fort Duquesne — that's where our great city of Pittsburgh now stands in Pennsylvania.

"Braddock wouldn't listen to Washington's advice: 'separate the troops; let each man fight singly, Indian-fashion, using trees and rocks for protection.' Braddock was ambushed and killed. Washington took over the command and managed to retreat the British soldiers to Fort Cumberland. Later, he commanded the fort.

"Many years passed. Washington won the independence of our country from yours, Mike. In 1794, as the first President of the United States, Washington returned to the familiar parade ground of Fort Cumberland. He was dressed for the last time in his splendid, bemedalled, uniform as the Commander-in-Chief of the Continental Army. He came to review the troops called out to suppress a rebellion, when the citizens of Pennsylvania objected to a tax on whiskey."

Captain Clark noted the excitement in Mike's eyes and realized that the lad was truly interested in the British-American history he was hearing. And the captain, too, was proud that he could still recite the facts he had long ago drilled into his own head and then been able to relate to his students.

"Only 11 years after our infant nation declared itself free of your mighty empire, Cumberland was established. America's first Great National Pike began here, as I've already told you. Between 1834 and 1845, it stretched westward from Baltimore — where you landed from England — over six states, ending in Vandalia, Illinois,

"That National Pike was once considered the finest macadamized road in the world. Several stagecoach lines operated regularly over the highway and Cumberland — like Hancock — became a bustling metropolis. Bright-painted coaches, drawn by high-spirited horses, rumbled over the road, careening around canvas-covered Conestoga wagons that lumbered along, laden with supplies for Western pioneers.

"However, when railroads appeared on the scene, stagecoach service between Baltimore and St. Louis stopped. Travel by rail over smooth tracks was far faster, less bumpy, more comfortable. That is, if you didn't mind the clouds of black cinders and smoke that enveloped the trains and covered the passengers with soot . . ."

The captain was silent for awhile. He reached down and rumpled Mike's hair, playfully. "You've still got that faraway look in your eyes, Mike, as though you're still back in Colonial times. Perhaps I've offended you with all that history about when your ancestors and mine were at each other's throats. If so — I'm sorry. Those family quarrels have faded into the past — like the feud between the blackeyed and blue-eyed Twiggs. We're still family — aren't we? You and I and our separate countries."

"Righto!" Mike brightened and grinned pleasantly up at his uncle. Then, with a wink — and to prove he had learned some American slang — doubtless from Cousin Abby, he added: "You bet! Okay. Gee whiz!"

Captain Clark roared. "Come along, now. I've filled your time with too much history and not enough work. Let's get back, first, to your aunt and cousin."

The Clarks, Mike, and old John spent that day and the next helping the dockhands unload the hold of the *Daisy Mae.* Then they visited George Washington's log cabin headquarters, built for him by Braddock's men. Mike signed the guest book proudly, entering "England" in the space for his address.

"George Washington's ancestors came from jolly old England — just like me, you know," he informed the old caretaker who had handed him the ledger. "And these are my American relatives," he said, introducing them.

On their last day in Cumberland, the crew of the *Daisy Mae* went shopping. When they returned, they saw a new Model T Ford motorcar at the curb of the Fort Cumberland Hotel entrance. Curious passersby were crowded around, admiring the gleaming brass headlamps and radiator. But the black hood was lifted from over the engine and the long-faced owner of the vehicle was bent over and peering inside. His head was almost hidden in the white cloud of steam that hissed from beneath the car and around the wooden spokes of the wheels.

One man turned away from the crowd. "Get a horse!" he said, roughly, to the sad-looking motorist dressed in linen duster, gloves, eye goggles, and cloth cap. Muttering, the man shuffled off, down the street.

"Sounds like he, too, has had some bad luck with horseless carriages, doesn't it?" commented Captain Clark. "Come," he said, "let's get our dinner and lodgings."

They enjoyed a hearty meal of beefstew, asparagus, and applesauce. For dessert, the adults had tea and deep dish cherry pie; Abby and Mike had strawberry ice cream and cookies.

The captain paid for clean, comfortable lodgings and led the way to the wire-caged elevator, which hoisted them to the top floor. In their rooms, they quickly took turns soaking in tubs of hot water.

Later, clean, refreshed, and pleasantly sleepy, Abby and Mike lay in extra cots set up in the bedroom. Captain and Mrs. Clark sat in the parlor, in comfortable horsehair-covered chairs, under electric lamps, reading *The Cumberland News* and catching up on events since they left Washington City.

Abby was staring up at the ceiling, where a crystal chandelier glittered in the light from its electric bulb. Her arms were folded on the pillow under her head.

"You know, Mike," she said, across the room. "I've been thinking what a nice fellow you are — even if you'll be growing up into an Englishman."

"And, you know, Abby," replied Mike, mockingly. "I've been thinking what a nice person *you* are — even if you'll be growing up into a female Yankee Doodle!"

Then he yawned and yawned again; he was half-awake but thinking of the return voyage to Georgetown — and of all the adventures still waiting for him along the old C & O Canal.

Mike mumbled to himself, memorizing the canal ditty John Manson had taught him. Mike planned to sing it, loud and clear, while driving one of the mules, as the *Daisy Mae* approached Widewater. That would be on the home stretch — about 13 miles from Washington City:

"I was coming down the Log Wall,
An' give a mighty yell —
The mules got scared and
 the driver catched hell!
The cap'n played the fiddle;
The steersman played the flute;
The cook poured coffee in the
 ol' man's boot . . .
Git along, ol' bones! Git along!"

And with a grin spreading over his face, Mike fell asleep.